Neuromorphic Computing Principles
and Organization

Abderazek Ben Abdallah • Khanh N. Dang

Neuromorphic Computing Principles and Organization

 Springer

Abderazek Ben Abdallah
School of Computer Science
University of Aizu
Aizu-Wakamatsu, Fukushima, Japan

Khanh N. Dang
School of Computer Science
University of Aizu
Aizu-Wakamatsu, Fukushima, Japan

ISBN 978-3-030-92527-7   ISBN 978-3-030-92525-3   (eBook)
https://doi.org/10.1007/978-3-030-92525-3

This Springer imprint is published by the registered company Springer Nature Switzerland AG
The registered company address is: Gewerbestrasse 11, 6330 Cham, Switzerland

To my parents, my wife, Sonia, and my children, Tesnim and Beyram

ABA

Dedicated to my parents

KND

Preface

As the end of Moore's law seems closer than ever, computer scientists have been exploring to build machines as complex and efficient as our brain, dealing with power density and clock frequency challenges of the conventional architecture. Our brain works entirely differently compared to traditional von Neumann architecture. There are many secrets behind how the human brain works. We know that it distributes computation and memory among more than 100 billion biological neurons, and each of them is connected with thousands of others via synapses. Neurons communicate with each other through spikes (i.e., short electrical pulses). The brain is a powerful computation system that helps us survive, adapt, and predict, while consuming tens of watts.

Brain-inspired or neuromorphic computing is a biologically inspired approach created from highly connected neurons to model neuroscience theories and solve machine learning problems. The term neuromorphic was first introduced by Carver Mead in 1990, where it referred to very-large-scale integration (VLSI) with analog components to mimic biological neural systems. Such systems can be categorized into non-spiking and spiking approaches. First, the non-spiking approach is referred to as the implementation of traditional artificial neural networks (ANNs) which aims to improve the throughput over the power consumption (or acceleration purpose). In recent years, ANNs have shown a remarkable improvement in terms of accuracy for large-scale visual/auditory recognition and classification tasks. Notably, the convolution neural network (CNN) and recurrent neural network (RNN) have shown to be promising tools for a wide range of applications such as image, video, and speech. They are typically trained by using graphic processing units (GPUs) or on the cloud side. The state-of-the-art neural networks tend to increase their number of layers and size (i.e., deep learning). However, this leads to challenges for hardware systems in terms of computation, memory, and communication resources.

The neuromorphic computing systems promises to drastically improve the efficiency of critical computational tasks such as decision making and perception. Unlike the typical artificial neural networks (ANNs), where neurons fire at each propagation cycle, the neurons in a brain-inspired neural networks model, named spiking neural networks (SNNs), fire only when a membrane potential reaches a specific value. Spiking neurons are only activated when sufficient signals are integrated from other neurons, which leads to sparse neural activities at the network level. Hence, the large spike sparsity and simple synaptic operations in the network enable SNNs to outperform ANNs in terms of energy efficiency.

This book stands independent and is organized into nine chapters. We have made every attempt to make each chapter self-contained. Chapter 1 introduces the neuromorphic computing system and explores the fundamental concepts of artificial neural networks. We first discuss biological neurons and the dynamics that are abstracted from them to model artificial neurons. Next, we discuss artificial neurons and how they have evolved in their representation of biological neuronal dynamics. Afterward, we discuss implementing these neural networks in terms of neuron models, storage technologies, inter-neuron communication networks, learning, and various design approaches.

Chapter 2 presents the fundamental design principle to build an efficient neuromorphic system in hardware. The challenges that need to be solved toward building in hardware a spiking neural network architecture (neuromorphic) with many synapses include building a small-sized massively parallel architecture with low-power consumption, efficient neuron coding scheme, and lightweight on-chip learning algorithm. The other major challenge is the on-chip communication and routing network, which allows data to be communicated between neurocores and off-chip data to be transferred to the cores. The constraints mentioned above make the deployment of such a brain-like IC a challenging on-chip interconnect problem.

Chapter 3 presents how learning in neuromorphic computing systems is conducted. Neuromorphic hardware's primary goal is to emulate brain-like neural networks to solve real-world problems. However, training on neuromorphic systems is challenging to the required non-local computations of gradient-based learning algorithms. Spiking neural networks gained popularity by incorporating learning. In these neural networks, there are two fundamental modes: Inference and learning. The learning phase, which minimizes a particular cost (loss) function, is a complex process of acquiring the parameters to output the correct inference results. In contrast, inference is computing the output values based on the given input and the network parameters.

To design a neuromorphic system on hardware, it is imperative to develop artificial neurons that mimic biological neurons and artificial synapses that emulate biological synapses. Recently, numerous efforts have been made to realize artificial synapses using post-CMOS devices, including resistive random access memory (ReRAM), ferroelectric field-effect transistor (FeFET), phase change memory devices, magnetoresistive random access memory (MRAM). A non-CMOS neuron based on emerging devices has also been investigated. Chapter 4 discusses the major emerging memory technologies that promise neuromorphic computing and highlight some recent significant progress on device studies. The advantages and challenges for each device technology are also discussed.

The brain connectivity is generally described at several levels of scale, including individual synaptic connections that link individual neurons at the microscale, networks connecting neuronal populations at the mesoscale, and brain regions linked by fiber pathways at the macroscale. Since each neuron is connected to many others, high bandwidth is required. Moreover, since the spike times are used to encode information, very low communication latency is also needed. Chapter 5 presents the circuits and architectures used for communication in neuromorphic systems. In particular, the Network-on-Chip fabric is introduced for receiving and transmitting spikes following the Address Event Representation (AER) protocol and the memory accessing method. First, the chapter describes the interconnect method for inter-neurons communication. Second, the interconnect design principle is covered to help understand the overall concept of on-chip and off-chip communication. The remaining parts cover advanced on-chip interconnect technologies, including si-photonic three-dimensional interconnects and fault-tolerant routing algorithms.

To develop such emerging systems, designers use large-scale models on dedicated hardware platforms, such as FPGAs, GPUs, or ASICs. The designers need a long time to collect datasets, train, and design accelerators to keep the trained models private and reliable. However, with the growing complexity of neuromorphic systems, there are severe vulnerabilities in the hardware implementations. An attacker who does not know the details of structures and designs inside these accelerators can effectively reverse engineer the neural networks by leveraging various side-channel information. Moreover, as neuromorphic systems are complex and integrate large number of neurons and synapses, the fault probability is accumulated and can threaten system reliability. Chapter 6 covers the main threats of reliability, and discusses several recovery methods.

Chapter 7 presents the architecture and hardware design of a reconfigurable spiking neuromorphic system. The architecture implements a Multi-Layer Percep-tron (MLP) that can be reconfigured to recover from faults with suitable methods that use an FPGA without being dependent on FPGA intellectual property (IP). This approach makes possible its implementation in application-specific integrated circuits (ASICs). Most spiking neuromorphic designs mainly focused on fixed functionality using available off-the-shelf components. Such an approach is lacking the flexibility to adapt to various computing environments. A reconfigurable design approach supports multiple target applications via dynamic reconfigurability, network topology independence, and network expandability.

Chapter 8 presents a real hardware-software design of a reliable three-dimensional digital neuromorphic processor geared explicitly toward the 3D-ICs biological brain's three-dimensional structure. The platform enables high integration density and slight spike delay of spiking networks and features a scalable design. R-NASH is a design based on the through-silicon-via (TSV) technology, facilitating spiking neural network implementation on clustered neurons based on network-on-chip (NoC). The system provides a memory interface with the host CPU, allowing for online training and inference of spiking neural networks. Moreover, R-NASH supports fault detection and recovery with graceful performance degradation.

Chapter 9 presents a comprehensive survey of the research of neuromorphic computing systems. First, the chapter gives the motivations of neuromorphic computing. Then, it describes significant research works in the field, which we categorize as software emulation approach, digital hardware approach, and analog and mixed-signal hardware approaches. This chapter aims to provide an exhaustive review of the research conducted in neuromorphic computing and illuminates the gaps in the field where new research is needed.

The neuromorphic computing principles and organization book is an excellent resource for researchers, scientists, graduate students, and hardware-software engineers dealing with the ever-increasing demands on fault-tolerance, scalability, and low power consumption. It is also an excellent resource for teaching advanced undergraduate and graduate students about the fundamentals concepts, organization, and actual hardware-software design of reliable neuromorphic systems with learning and fault-tolerance capabilities.

Acknowledgments

This book took nearly 3 years to complete. It evolved from our research and education experiences in adaptive computing systems and neuromorphic computing architectures designs. The Neuromorphic computing paradigm created excellent opportunities to explore cognitive AI system performance and created many design challenges that designers must overcome. To advance the field of neuromorphic computing, the exploration of novel materials and devices will be the key to improve the power efficiency and scalability of state-of-the-art CMOS solutions. Thus, we must continue innovating new algorithms and techniques to solve these challenges. We must also educate computer science and computer engineering students in both neuromorphic computing and engineering. The authors wish to thank Mark Ogbodo, Zhishang Wang, and Wang Jiangkun from the Adaptive Systems Laboratory at the University of Aizu, and Vu Huy The for their valuable comments, help, and discussion.

Finally, this first version of this book was completed without describing the didactic materials pedagogically as expected in a textbook with exercises and their solutions at the end of each chapter. Hopefully, those goals will be completed in the second edition of this book after receiving insightful feedback from students, instructors, researchers, and practicing engineers. We truly appreciate it if you give us such feedback, allowing us to prepare a second edition for this fast-growing and emerging computing paradigm.

Aizu-Wakamatsu, Japan   Abderazek Ben Abdallah
Aizu-Wakamatsu, Japan   Khanh N. Dang

Contents

Acronyms

AAAI	American Association for Artificial Intelligence
ACK	Acknowledgment
ACM	Association for Computing Machinery
ADC	Analog to Digital Converter
ADM	Approximate Derivative Method
AER	Address Event Representation
AHB	Advanced High-performance Bus
AI	Artificial Intelligence
AMBA	Advanced Microcontroller Bus Architecture
AMD	Advanced Micro Devices
ANN	Artificial Neural Network
API	Application Programming Interface
ARM	Advanced RISC Machines
ASIC	Application-Specific Integrated Circuit
ASID	Anti-Counterfeiting, Security and Identification
BCM	Building Cube Method
BE	Best Effort
BL	Bit Line
BP	Backpropagation
BW	Buffer Writing
BWCCA	Broadband and Wireless Computing, Communication and Applications
CAD	Computer Aided Design
CAM	Content-Access-Memory
CASES	Conference on Compilers, Architecture and Synthesis for Embedded Systems
CD	Compact Disc
CICC	Custom Integrated Circuits Conference
CIFAR	Canadian Institute For Advanced Research
CLEO	Conference on Lasers and Electro-Optics
CMD	Command

CMOS	Complementary Metal-Oxide Semiconductor
CNN	Convolution Neural Network
CP	Configuration Packet
CPU	Central Processing Unit
CRC	Cyclic Redundancy Code
CS	Computer Science
CS	Chip Select
CSUR	Computing Surveys
CT	Crossbar Traversal
DAC	Digital-to-Analog Converter
DBN	Deep Belief Network
DMA	Direct Memory Access
DNN	Deep Neural Network
DOR	Dimension Order Routing
DRAM	Dynamic Random Access Memory
DTCM	Data Tightly Coupled Memory
DVD	Digital Versatile Disk
DWDM	Dense Wavelength Division Multiplexing
DWM	Domain Wall Memory
EC	Electronic Controller
ECC	Error Correction Code
ECN	Electronic Control Network
ECTC	Electronic Components and Technology Conference
EDRAM	Embedded Dynamic Random Access Memory
EMCSI	Electromagnetic Compatibility Signal/Power Integrity
ESD	ElectroStatic Discharge
ETE	End-To-End
FeFET	Ferroelectric Field-Effect Transistor
FIFO	First In, First Out
FPGA	Field Programmable Gate Array
FPNA	Field Programmable Neural Array
FT-PHENIC	Fault-Tolerant Photonic Network-on-Chip
FTMC-3DR	Fault-Tolerant Multicast 3D Routers
FTPP	Fault-Tolerant Photonic Path-configuration algorithm
FTSP-KMCR	Fault-Tolerant Shortest Path K-means-based MultiCast Routing algorithm
FTSPKMCR	Fault-Tolerant MultiCast Routing algorithm
GA	Genetic Algorithm
GPU	Graphic Processing Units
GPGPU	General Purpose Graphic Processing Units
GS	Greedy Search
GUI	Graphical User Interface
HAL	Hardware Abstraction Layer
HDD	Hard Disk Drive
HiPC	High Performance Computing

HPCA	High Performance Computer Architecture
HRS	High Resistive state
IBM	International Business Machines
IC	Integrated Circuit
ID	Identification
IEEE	Institute of Electrical and Electronics Engineers
IF	Integrate and Fire
ILP	Integer Linear Programming
IP	Intellectual Property
i-PACT	Innovations in Power and Advanced Computing Technologies
ISBN	International Standard Book Number
ISCA	International Symposium on Computer Architecture
ISCAS	International Symposium on Circuits and Systems
ISI	Inter-Spike-Interval
ITCM	Instruction Tightly Coupled Memory
JPEG	Joint Photographic Experts Group
KB	Kilobyte
KL	Kernighan-Lin
LA-XYZ	Look-Ahead XYZ
LIF	Leaky Integrate-and-Fire
LRS	Low-Resistive State
LRU	Least Recently Used
LTD	Long-Term Depression
LTP	Long-Term Potentiation
LUT	Look-Up-Table
MB	Mega Bytes
MCSoC	Multicore/Many-Core Systems-on-Chip
MFMC	Max-Flow Min-Cut
MIVs	Monolithic Intertier Vias
MJT	Multi Junction Technology
MLP	Multi-Layer Perceptron
MNIST	Modified National Institute of Standards and Technology database
MPI	Message Passing Interface
MR	Microring Resonator
MRAM	Magnetoresistive Random-Access Memory
MRCT	Micro Ring Configuration Table
MRPR	Microring fault-Resilient Photonic Router
MRST	Micro Ring State Table
MTBF	Mean Time Between Failures
MTJ	Magnetic Tunneling Junction
MTTF	Mean Time to Failures
MTTR	Mean Time to Repair
NACK	Negative Acknowledgment
NAND	Not and
NASH	Neuro-inspired ArchitectureS in Hardware

NEWS	North-East-West-South
NI	Network Interface
NP-hard	Non-deterministic Polynomial-time hard
NSEW	North South East West
NVM	Non-Volatile Memories
OE	Output Enable
OSI	Open Systems Interconnection
OSI	Open Systems Interconnection Model
PB	Path_Blocked
PCB	Process Control Block
PCM	Phase Change Memory
PCN	Photonic Communication Network
PE	Processing Element
PJ	Picojoule
PNoC	Photonic Network-on-Chip
PS	Photonic Switch
PSCP	Path-Setup-Control Packet
PSO	Particle Swarm Optimization
PSP	Post-Synaptic Potential
PV	Process Variation
RAM	Random-Access Memory
RBL	Resistance Between Layers
RC	Routing Calculation
RE	Read Enable
RELU	Rectified Linear Unit
RISC	Reduced Instruction Set Computing
RMP	Residual Membrane Potential
R-NASH	Reconfigurable Neuro-inspired ArchitectureS in Hardware
RNN	Recurrent Neural Network
ROM	Read-Only Memory
RPM	Randomized Partially Minimal
RRAM	Resistive Random-Access Memory
RSP	Rapid System Prototyping
RTL	Register-Transfer Level
SA	Switch Allocation
SAF	Store-And-Forward
SDRAM	Synchronous Dynamic Random Access Memory
SDSP	Spike Driven Synaptic Plasticity
SECDED	Single Error Correction, Double Error Detection
SEQ	Sequencer
SET	Single-Event Transients
SEU	Single-Event Upsets
SMC	Systems, Man and Cybernetics
SNN	Spiking Neural Network
SNPC	Spiking Neuron Processing Core

SoC	System On a Chip
SP	Shortest Path
SRAM	Static Random Access Memory
SRDS	Symposium on Reliable Distributed Systems
STA	Sciences and Techniques of Automatic Control and Computer Engineering
STDP	Spike Timing Synaptic Plasticity
STPD	Spike-Timing-Dependent-Plasticity
STT-RAM	Spin-Transfer Torque RAM
TDM	Time-Division Multiplexing
TDMA	Time-Division-Multiple-Access
TECS	Transactions on Embedded Computing Systems
TMR	Triple Modular Redundancy
TODAES	Transactions on Design Automation of Electronic Systems
TSVs	Through Silicon Vias
TTFS	Time-To-First-Spike
TV	Thermal Variations
UI	User Interface
USA	United States of America
USB	Universal Serial Bus
VCSEL	Vertically Cavity Surface Emitting Laser
VGG	Visual Geometry Group Network
VLSI	Very-Large-Scale Integration
VTS	VLSI Test Symposium
WDM	Wavelength Division Multiplexing
WE	Write Enable
WL	Word Line
WL	Worst Loss
WTA	Winner-Take-All
XOR	Exclusive OR

Chapter 1
Introduction to Neuromorphic Computing Systems

Abstract The term neuromorphic is generally used to describe analog, digital, mixed-mode analog/digital VLSI, and software systems that implement several models of neural systems. The implementation of neuromorphic computing on the hardware level can be realized by various technologies, including spintronic memories, threshold switches, CMOS transistors, and oxide-based memristors. This chapter introduces the neuromorphic computing systems and explores the fundamental concepts underlying this emerging paradigm. We first discuss biological neurons and the dynamics that are abstracted from them to model artificial neurons. Next, we discuss artificial neurons and how they have evolved in their representation of biological neuronal dynamics. Afterward, we discuss implementing these neural networks in terms of neuron models, storage technologies, inter-neuron communication networks, and learning.

1.1 Introduction

The human nervous system is composed of more than 100 billion cells known as neurons. The neurons perceive changes in the environment, convey these changes to other neurons, and directs body responses to these perceptions. Because these neurons can carry out information processing in a rapid, parallel, fault-tolerant, and energy-efficient manner, it has received so much attention. This chapter introduces the fundamentals of neuromorphic computing systems by discussing biological neurons and the dynamics abstracted from them to model artificial neurons.

Neuromorphic computing brain-inspired computing paradigm takes inspiration from the brain to develop energy-efficient circuits and systems for future information processing, capable of highly complicated tasks. Such computing promises to drastically improve the efficiency of critical computational tasks, such as decision-making and perception. Unlike the typical artificial neural networks (ANNs), where neurons fire at each propagation cycle, the neurons in a brain-inspired neural network model, named spiking neural networks (SNNs), fire only when a membrane potential crosses a threshold value. Spiking neurons are only activated

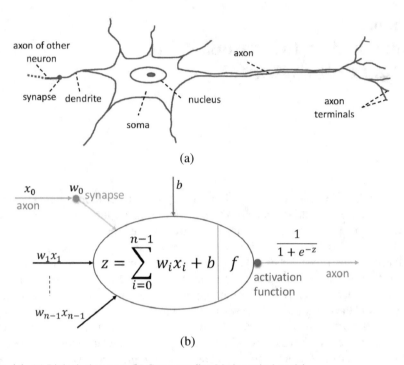

Fig. 1.1 (a) Biological neuron. (b) Corresponding Mathematical model

when sufficient signals are integrated from other neurons, which leads to sparse neural activities at the network level.

A general description of a biological neuron is presented in Fig. 1.1. A neuron consists of several parts: The dendrite, the axon, and the soma. The dendrite serves as an input channel to the neuron, while the axon serves as the output channel. A neuron receives electrochemical inputs from other neurons at the dendrites. Suppose the sum of these inputs is sufficiently powerful to activate the neuron. In that case, it transmits an electrochemical signal through the axon to other neurons whose dendrites are connected to any of its axon terminals. This connection among neurons is enabled via the synapses. The neuro-biological system is formidably connected. A typical cortical neuron has up to 10K inputs, and some cerebellar neurons have up to a quarter of a million inputs. Therefore, artificial neurons are designed to operate in a manner analogous to biological neurons.

Figure 1.1b shows a computational neuron model. The input signals (e.g., x_0) received from the axon of other neurons are multiplied with the weight of the synapse that connects them (e.g., w_0). The dendrite then transports weighted inputs (e.g., $w_0 x_0$) to the soma of the receiving neuron. The weighted inputs are summed up as the neuron membrane potential and passed through an activation function that maps it to the neuron's output (Fig. 1.2).

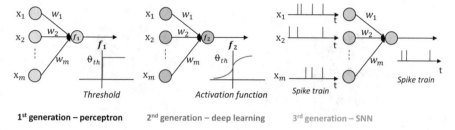

Fig. 1.2 Neural network generations

Over the years, several neural network modeling approaches have been proposed, differing in topology and features, to capture the dynamics of neural computation. These modeling approaches have evolved through three generations, keeping in mind the computational principles of the biological brain. In the first generation, the neurons were referred to as perceptrons. These perceptrons process only digital signals using a single layer. The sum of weighted inputs of this neuron is mapped to the neuron output using a binary threshold. Some examples of perceptrons include Hopfield networks and Boltzmann machines. The second generation, neurons, are called a conventional artificial neural network. It maps the sum of weighted inputs to the neuron output using activation functions such as sigmoid, exponential, and polynomial, which have a continuous set of possible outputs. Also, this second-generation network employs learning algorithms based on gradient descent. Examples of this generation of the neural network include Feedforward, radial basis function units, and recurrent sigmoid neural network. The third generation, referred to as the Spiking Neural Network (SNN), is modeled more analogous to the dynamics of biological neurons than previous generations. It is event-driven and operates by accumulating input spikes at its membrane potential. An output spike is fired by the neuron only when its membrane potential exceeds a certain threshold. To perform tasks, artificial neurons like their biological counterparts need to be connected. The manner in which neurons are connected determines their topology. A summary of some neural network topologies illustrated in Fig. 1.3 are described as:

- **Feed-forward neural networks (FFNN)**: This network topology described in Fig. 1.3 is organized into three categories of layers: The input, hidden, and output layer. The connections between neurons in this network are made across layers and not within a layer. Information flows in a forward direction from the input layer, through the hidden layer(s), and finally to the output layer. The backpropagation learning method is usually employed in training this network. The multilayer perceptron is another network with a similar topology as the feed-forward neural network. An example of a feed-forward neural network usually employed in pattern recognition and classification tasks is the radial basis function.

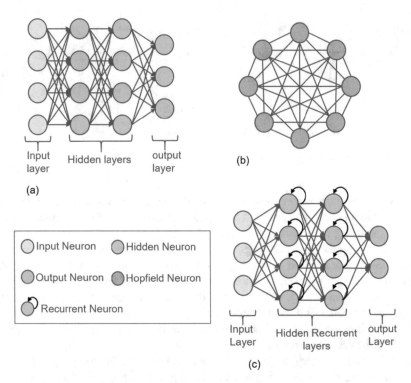

Fig. 1.3 Some prevalent neural network topologies. (**a**) Feedforward neural network. (**b**) Hopfield neural network. (**c**) Recurrent neural network

- **Hopfield neural network (HFs):** This network topology possesses cyclic and recursive characteristics. They are made up of binary threshold neurons with recurrent connections between them and can behave in several ways: settling in a stable state, oscillating, or following less predictable disorganized trajectories. Its global energy is determined by summing up several contributions, and each contribution can also be determined from one symmetric connection between neurons and the binary states of the two neurons.
- **Recurrent neural networks (RNNs):** The recurrent neural network is derived from the FFNNs. However, as described in Fig. 1.3c, its hidden layers are replaced with recurrent layers. The layers of an RNN receive inputs from previous layers and the output of its layer. The ability of RNNs to process sequences of inputs with their internal state makes them suitable for speech recognition and connected handwriting recognition.

1.2   Design Challenges

With the increasing demand for computing machines analogous to the biological brain, neuro-inspired computing has advanced to the exploration of neuro-inspired architectures that best the limitations of the traditional computer systems. Conventional computer systems are based on the Von Neumann architecture. However, the biological brain shows a disparity in structure, power consumption, and computational power compared to the traditional computer system. Therefore, a biologically inspired approach designed from densely connected neurons enables neuroscience theories to be modeled and machine learning problems to be solved.

Recent advances in artificial neural networks (ANN) have enabled machine learning tasks like visual/auditory recognition and classification to be performed on a large scale [28, 34, 37]. These ANNs, especially the convolutional neural network (CNN), have demonstrated exceptional performance in these tasks. However, as ANNs increase in size to enable them to perform more complex tasks, they inculcate more layers, which increases the hardware resources (power consumption and processing) required to simulate them. To mitigate the cost of simulating these ANNs, a spiking neural network approach was proposed. By mimicking the behavior of biological neurons more closely, SNNs can demonstrate low-power consumption in magnitudes of picojoule (pJ) compared to ANN [22].

Despite being a flexible way of exploring the behavior of neuronal systems, simulating large-scale SNN in software is slow and does not fully harness the energy efficiency of SNN. As an alternative, scalable hardware multicore neuromorphic architectures that can support a massive number of neurons and synapses, and leverage the parallelism and spike sparsity available in SNN to deliver rapid processing with low power is imperative. However, realizing such a neuromorphic architecture requires building small-sized spiking neuron cores with low-power consumption, an efficient neuron coding scheme, and a light-weight learning algorithm. These neurocores also need a scalable interneuron communication architecture that can manage the enormous amount of traffic generated by the massive number of neurons they embed. Given that communication architectures, such as shared buses, would perform poorly with the increased number of neurons, and two-dimensional packet-switched network-on-chip make it challenging to realize a high level of parallelism, moving to 3D integrated circuits (3D-ICs) interconnect is a suitable approach that allows for scalable designs with a high level of parallelism, shorter connections, and lower power consumption. However, highly dense neuromorphic architectures also encounter the reliability issue where a single point of failure can affect operation. Because neuromorphic systems rely heavily on spike communication, an interruption or violation in the timing of spike communication can adversely affect the performance and accuracy of a neuromorphic system. Therefore, adaptivity is required in neuromorphic systems to enable them to mitigate the effect of faults.

1.3 Neural Networks

1.3.1 Artificial Neural Networks

In conventional ANN, input signals and weights are represented with real values, and the implementation methods can be categorized either as analog or digital. Analog implementations offer rapid processing, power efficiency, and low area cost. However, they are susceptible to noise which makes representation of real numbers difficult and accuracy limited. Digital implementation, on the other hand, provides programmability, high precision, and reliability, but compared to analog performance, they suffer from high area cost and latency.

Learning in neural networks is simply the process of finding the best set of synaptic weights for maximizing a neural network's accuracy. Implementing learning has been one of the significant challenges in the design of neuromorphic systems. Learning in a neuromorphic system is implemented either on-chip or off-chip. The choice of implementation approach is made on many factors that include the neural network model, and hardware resources. Two of the notable neural network learning approaches are supervised and unsupervised learning. This subsection reviews those approaches and learning rules employed in training conventional artificial neural networks.

The learning approach entails training a neural network based on input-output pairs, where the network learns by example. One of the prevalent learning rules used in training conventional ANN is the backpropagation (BP) learning rule. This learning rule trains an ANN by modifying its synaptic weights based on the error rate obtained in its previous training stage. It can be employed in training neural network topologies such as feed-forward, recurrent, and convolution. The BP learning rule is usually implemented off-chip [27, 48] on a traditional host machine. There it is used to train ANN weights, and after the training, the trained weights are mapped to the neuromorphic chip. This learning rule has demonstrated high precision in its training, taking advantage of the software platform. However, it is not suitable for neuromorphic systems that require frequent re-training of their synaptic weights. Several on-chip implementations of BP on neuromorphic systems [5, 8], and variants of it, have also been either optimized or simplified for on-chip implementation [12, 15]. Other supervised learning algorithms include support vector machines and linear regression.

In contrast to the supervised learning approach, the unsupervised learning approach has no input-output pair and has no example to learn from. This learning approach is quite unpopular, and there have been some unsupervised learning rules based on self-organizing maps [6, 35, 46] that have been implemented on neuromorphic chips.

1.3.2 Spiking Neural Networks

Spiking neural networks (SNNs) are artificial neural network models that more closely mimic biological neural networks. In addition to neuronal and synaptic states, SNNs incorporate variant time scale into their computational model. Since each neuron in these networks is connected to thousands of others, high bandwidth is required [49, 50]. Moreover, since the spike times are used to encode information in SNN, very low communication latency is also needed.

Figure 1.4a shows the biological neuron model where dendrites receive inputs from upstream neurons via the synapses. Incoming spikes are integrated into soma as its membrane potential. If the membrane potential crosses the threshold, the neuron issues an outgoing spike to an axon. The axon sends the spike to the downstream neurons via synapses. Figure 1.4b shows an example of a spiking neuron. The input spikes are multiplied (memory reading) with the corresponding weights to have weighted inputs. The membrane potential is accumulated from the weighted inputs and creates an outgoing spike if it is higher than the threshold. In the spiking neuron models, there are three major parameters for being stored: *(1)* incoming spikes; *(2)* synaptic weights, and *(3)* neuron's internal parameters (membrane potential, threshold, etc.).

In SNN, biological neurons communicate information via short electrical pulses referred to as spikes. Neural coding focuses on how this information is represented with electrical activity both at the neuron level and in networks of neurons. Efforts have been made over the years to determine how these spikes can be encoded to contain information. Some coding schemes that have been proposed are presented in [24]. The rate coding scheme, which is sometimes called frequency coding, conveys information based on a neuron's firing rate, proportional to the stimulus level. Several studies have shown that the temporal resolution of neural code is on a scale of milliseconds. Therefore in temporal coding, information is represented in the precise timing of spikes. As described in previous sections, SNN, unlike its predecessors, mimics the brain's behavior more closely, and over the years, several spiking neuron models have been proposed. These spiking neuron models [31] differ in the level of details they abstract from biological neurons. In this subsection, we describe some of the prevalent models.

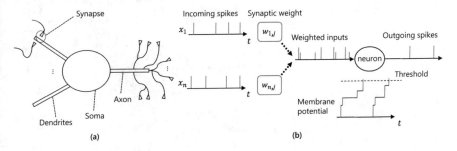

Fig. 1.4 (a) Biological neuron. (b) Spiking neuron

The Hodgkin-Huxley neuron model, which was proposed in the early 1950s [29], is a mathematical representation of neuron dynamics. It presents a mathematical description of the electric current through the membrane potential, giving the details of spike generation. The Hodgkin-Huxley model is the most biological plausible; however, its complexity with many parameters consumes a tremendous amount of hardware resources making it costly for large-scale implementations. Compared to Hodgkin-Huxley, a less complex model was proposed by Izhikevich [30]. This model computes a broad range of neuron spiking patterns using mathematical equations.

In summary, among the existing spiking models, Hodgkin-Huxley [29], Izhike-vich [30], and Leaky Integrate-and-Fire (LIF) [11] are often used. The Hodgkin-Huxley model is the best when measurable physiological parameters are highly considered. However, it consists of many coefficients, which leads to high cost when implementing large SNNs. In contrast, we can simulate hundreds of thousands of neurons when using LIF neural model; but it is incapable of producing rich spiking patterns. Finally, Izhikevich exhibits a good compromise in terms of biophysical similarity and computational cost. It is close to the Hodgkin-Huxley model in biological plausibility while analogous to the LIF in computational complexity.

1.4   Learning in Spiking Neural Networks

Spike timing-dependent plasticity (STDP) is the most popular learning rule imple-mented in neuromorphic systems [16, 32, 51]. It is Hebbian-based, coming from observation in the biological brain [44]. The operation of STDP depends on the arrival time of presynaptic spikes, in which the synaptic weight will be increased when the spike arrives before the post-synaptic neuron "fire" and vice versa, as illustrated in Fig. 1.5. It is an unsupervised learning rule and generally implemented on-chip thanks to its friendly hardware resource.

Fig. 1.5 STDP architecture

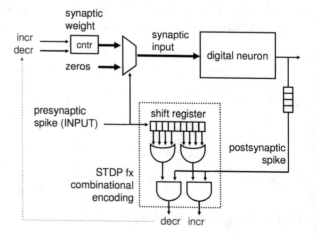

Apart from the majority of STDP, spiking neuromorphic system also adopted supervised learning rules. In this case, such systems use a "teacher" signal during the training phase. Besides, another work [21] successfully implemented the spike-driven synaptic plasticity (SDSP) learning rule. Unlike STDP, this learning rule induces an update each time a pre-synaptic spike occurs. On the other hand, backpropagation is also adopted for spiking systems [52]. In [19], authors first train an ANN with BP, then convert it into SNN by mapping real-value inputs/activations to average firing rates of Poisson spikes. This mechanism can be adopted to implement on spiking hardware as an off-chip learning method.

1.5 Synapse Memory Technologies

In traditional computer systems, memory speed is a bottleneck because processors have significantly improved over the years, surpassing memory speed and throughput. This leaves processors idle while waiting for memory. But SNN is different. Its architecture provides memory in company with processing; they both operate in parallel. These spiking neurons process with events, and as opposed to traditional processors that work and communicate in megahertz and gigahertz ranges, they communicate in around 10 Hz. This difference in speed enables neuromorphic processors to use time multiplexing to combine many events into a single communication channel. In SNN, storing and reading synaptic weights constitutes the primary operation, and designing a large SNN with an enormous number of synapses will require large memory bandwidth. So while communication speed is not somewhat a challenge in neuromorphic systems, memory bandwidth needs to be overcome. Researchers have taken several approaches to realize these synapses in CMOS by exploring various memory technologies in addressing this challenge. Some of these memory technologies include; static random-access memory (SRAM), a prominent memory technology in semiconductor design. In [43] a neuromorphic chip with 256 neurons implements a transposable 8-transistor SRAM-based which grants row and column access. A typical SRAM contains six transistors (6-T), and although having high leakage current and low density, SRAM offers multiple read and write. Its significant advantages are speed and reliability when compared to other memory technologies [9].

The two more transistors on the 8-T SRAM adds access to the word and bit lines in transposed orientation to the typical 6-T design. To handle the general leakage power of the chip, the authors leveraged ultra-high-Vt devices which reduced it by 3, at the cost of the increased minimum operating voltage of the memory array. Another memory technology is embedded dynamic random-access memory (eDRAM), which was used in [33] to design a high-density 3D memory for a programmable digital neuromorphic architecture. eDRAM is a capacitor-based memory integrated on the same multi-chip module. Due to its simple conventional one 1-transistor 1-capacitor design, it is inclined to having less area cost when compared to SRAM. However, with the gradual leakage of its storage charge during operation, its retention period is low. Also, its design is not easily compatible with CMOS. In [47], a spin-transfer torque ram (STT-RAM) was proposed as a stochastic indexmemristivememristive synapse for neuromorphic systems. An STT-RAM is a magnetic RAM that uses magnetic tunneling junction (MTJ) in its cells, and its simple design has a small area compared to SRAM and eDRAM. By adopting a schema proposed in [45], the authors organized the STT-RAM MJT as a crossbar connecting input and output neurons. However, the magnetization of the STT-RAM makes the writing process slow and consumes more energy. Another memory technology that has been used in the design of neuromorphic chips is resistive random-access memory (RRAM or ReRAM). It is a memory technology that relies on the resistance change of its cells to store information, and it has similar architecture to eDRAM. Many FPGA-based hardware emulators for neuromorphic chips with RRAM-based crossbar were proposed. RRAM offers low area, low power, and easy integration on CMOS. However, it suffers from stuck at faults (SAF) [54], which may cause short-circuiting and lead to increased power and dynamic switching variation large variation in the resistance of the memory. Other memory technologies include Phase Change Memory (PCM).

1.6 Neurons Communication Network

Communication architectures for spiking neuromorphic systems are responsible for delivering spikes between neuro-cores/tiles. They can be categorized as intra-chip and inter-chip. For inter-chip, address event presentation (AER) is commonly employed [10, 39]. In AER, each neuron has a unique address. Whenever a neuron generates a spike, its address is sent to post-synaptic neurons by a high-speed digital bus. AER is suitable for SNN implementations since it only needs to be active whenever neurons fire. To scale up the system, a hierarchical AER as a tree structure can be implemented.

On the other hand, network-on-chip (NoC) is commonly implemented for on-chip communication [1–3]. In the early stage of neuromorphic system implementations, buses are employed in some systems [40]. However, when comparing bus, tree, point to point, and mesh-based system, the results show that mesh with multicast offers the highest performance for SNN implementations. Later systems adopted mesh-based interconnect as the common NoC topology [4, 17]. In SpiNNaker [23], toroid NoC is used for forming a hybrid on-chip and off-chip interconnect system. Furthermore, AER also is used for on-chip communication [4, 17, 18, 23].

1.7 Neuromorphic System Design Domains

Full custom digital ASIC has been common platforms for spiking neuromorphic implementations. Two well-known examples of this kind of implementation are TrueNorth [4] and SpiNNaker [23]. While TrueNorth only supports the leaky integrate and fire neuron model with no on-chip learning, SpiNNaker offers extreme flexibility in the neuron model, synaptic model, and learning algorithm. However, TrueNorth has a benefit of energy efficiency by consuming 25 pJ per connection, while the SpiNNaker is 10 nJ per connection, as reported in [22]. Also, FPGAs are commonly used for implementing spiking neuromorphic systems [14, 25, 38]. They can be implemented as a part of the system and also as final implementations. While FPGAs are considered to be an excellent choice for acceleration over software simulations, they are not targeted as platforms for achieving low power.

Some characteristics that make analog platforms suitable for spiking implementations include: conservation of charge, amplification, thresholding, and integration. Therefore, there are a large number of implementations [41]. Besides, analog platforms are also designed to operate in subthreshold [53], and superthreshold modes for the speed-up purpose [42]. On the other hand, field-programmable analog arrays (FPAAs) have been used as other analog platforms. They are also customized for neural network implementation such as field programmable neural array (FPNA) [20] and Neuro FPAA [36] where they provide programmable components such as neurons, synapses.

Mixed-signal designs are also standard for neuromorphic systems [26] to take advantage of both analog and digital platforms. In these works, weights or other parameters are stored in digital memories to enable the system to be less noisy and more reliable. Furthermore, inter-chip and intra-chip communication architectures are also implemented in digital [13]. On the other hand, neurons are generally in the form of analog. Two well-known systems for this kind of implementation are Neurogrid [7] and BrainScales [42].

1.8 Chapter Summary

This chapter presented an overview of artificial neural networks, including spiking neural networks as the latest generation and how they are implemented. Spiking Neural networks can simulate biological neural networks with extreme energy efficiency thanks to event-based operations and fewer operation computations. The next chapter presents the interconnection network and how they solve the interconnect challenge in spiking neuromorphic systems.

References

1. Abdallah AB (2017) Advanced multicore systems-on-chip: architecture on-chip network, design. Springer, Berlin
2. Ahmed AB, Abdallah AB (2013) Architecture and design of high-throughput, low-latency, and fault-tolerant routing algorithm for 3d-network-on-chip (3d-noc). J Supercomput 66(3):1507–1532

3. Ahmed AB, Abdallah AB (2014) Graceful deadlock-free fault-tolerant routing algorithm for 3d network-on-chip architectures. J Parallel Distrib Comput 74(4):2229–2240.
4. Akopyan F, Sawada J, Cassidy A, Alvarez-Icaza R, Arthur J, Merolla P, Imam N, Nakamura Y, Datta P, Nam G, Taba B, Beakes M, Brezzo B, Kuang JB, Manohar R, Risk WP, Jackson B, Modha DS (2015) Truenorth: design and tool flow of a 65 mW 1 million neuron programmable neurosynaptic chip. IEEE Trans Comput Aided Des Integr Circuits Syst 34(10):1537–1557
5. Bayraktaroglu I, Ogrenci AS, Dundar G, Balkir S, Alpaydin E (1997) Annsys (an analog neural network synthesis system). In: Proceedings of international conference on neural networks (ICNN'97), vol 2, pp 910–915
6. Ben Khalifa K, Girau B, Alexandre F, Bedoui MH (2004) Parallel FPGA implementation of self-organizing maps. In: Proceedings of the 16th international conference on microelectronics, 2004. ICM 2004, pp 709–712
7. Benjamin BV, Gao P, McQuinn E, Choudhary S, Chandrasekaran AR, Bussat JM, Alvarez-Icaza R, Arthur JV, Merolla PA, Boahen K (2014) Neurogrid: a mixed-analog-digital multichip system for large-scale neural simulations. Proc IEEE 102(5):699–716
8. Berg Y, Sigvartsen RL, Lande TS, Abusland A (1996) An analog feed-forward neural network with on-chip learning. Analog Integr Circuits Signal Process 9(1):65–75
9. Bhaskar A (2017) Design and analysis of low power SRAM cells. In: 2017 Innovations in power and advanced computing technologies (i-PACT), pp 1–5
10. Boahen KA (1998) Communicating neuronal ensembles between neuromorphic chips. Springer US, Boston, MA, pp 229–259
11. Burkitt N (2006) A review of the integrate-and-fire neuron model: I. homogeneous synaptic input. Biol Cybern 95(1):1–19
12. Carvajal G, Figueroa M, Sbarbaro D, Valenzuela W (2011) Analysis and compensation of the effects of analog VLSI arithmetic on the LMS algorithm. IEEE Trans Neural Netw 22(7):1046–1060
13. Charles G, Gordon C, Alexander WE (2008) An implementation of a biological neural model using analog-digital integrated circuits. In: 2008 IEEE international behavioral modeling and simulation workshop, pp 78–83
14. Cheung K, Schultz SR, Luk W (2012) A large-scale spiking neural network accelerator for FPGA systems. In: International conference on artificial neural networks. Springer, Berlin, pp 113–120
15. Choi M, Salam FMA (1991) Implementation of feedforward artificial neural nets with learning using standard CMOS VLSI technology. In: IEEE international symposium on circuits and systems 1991, vol 3, pp 1509–1512
16. Dan Y, Ming Poo M (2004) Spike timing-dependent plasticity of neural circuits. Neuron 44(1):23–30
17. Davies M et al (2018) Loihi: a neuromorphic manycore processor with on-chip learning. IEEE Micro 38(1):82–99
18. Deiss SR, Douglas RJ, Whatley AM (1999) Pulsed neural networks. In: A pulse-coded communications infrastructure for neuromorphic systems. MIT Press, Cambridge, MA, pp 157–178
19. Diehl PU, Neil D, Binas J, Cook M, Liu S, Pfeiffer M (2015) Fast-classifying, high-accuracy spiking deep networks through weight and threshold balancing. In: 2015 International joint conference on neural networks (IJCNN), pp 1–8
20. Farquhar E, Gordon C, Hasler P (2006) A field programmable neural array. In: 2006 IEEE international symposium on circuits and systems, p 4117
21. Frenkel C, Lefebvre M, Legat J, Bol D (2019) A 0.086-mm^2 12.7-pj/sop 64k-synapse 256-neuron online-learning digital spiking neuromorphic processor in 28-nm CMOS. IEEE Trans Biomed Circuits Syst 13(1):145–158
22. Furber S (2016) Large-scale neuromorphic computing systems. J Neural Eng 13(5):051001
23. Furber SB, Galluppi F, Temple S, Plana LA (2014) The spinnaker project. Proc IEEE 102(5):652–665

24. Gerstner W, Kistler W (2002) Spiking neuron models: single neurons, populations, plasticity. Cambridge University Press, Cambridge
25. Glackin B, McGinnity TM, Maguire LP, Wu Q, Belatreche A (2005) A novel approach for the implementation of large scale spiking neural networks on FPGA hardware. In: International work-conference on artificial neural networks. Springer, Berlin, pp 552–563
26. Hahnloser RHR, Sarpeshkar R, Mahowald MA, Douglas RJ, Seung HS (2000) Digital selection and analogue amplification coexist in a cortex-inspired silicon circuit. Nature 405(6789):947–951
27. Haykin S (1998) Neural networks: a comprehensive foundation, 2nd edn. Prentice Hall PTR, Upper Saddle River, NJ
28. He K, Zhang X, Ren S, Sun J (2016) Deep residual learning for image recognition. In: Proceedings of the IEEE conference on computer vision and pattern recognition, pp 770–778
29. Hodgkin A, Huxley A (1952) A quantitative description of membrane current and its application to conduction and excitation in nerve. J Physiol 117:500–544
30. Izhikevich EM (2003) Simple model of spiking neurons. IEEE Trans Neural Netw 14(6):1569–1572
31. Izhikevich EM (2004) Which model to use for cortical spiking neurons? IEEE Trans Neural Netw 15(5):1063–1070
32. Jin X, Rast A, Galluppi F, Davies S, Furber S (2010) Implementing spike-timing-dependent plasticity on spinnaker neuromorphic hardware. In: The 2010 international joint conference on neural networks (IJCNN), pp 1–8
33. Kim D, Kung J, Chai S, Yalamanchili S, Mukhopadhyay S (2016) Neurocube: a programmable digital neuromorphic architecture with high-density 3d memory. In: 2016 ACM/IEEE 43rd annual international symposium on computer architecture (ISCA), pp 380–392
34. Krizhevsky A, Sutskever I, Hinton GE (2012) Imagenet classification with deep convolutional neural networks. Adv Neural Inf Process Syst 25:1097–1105
35. Kumar S, Forward K, Palaniswami M (1996) Performance evaluation of a RISC neuroprocessor for neural networks. In: Proceedings of 3rd international conference on high performance computing (HiPC), pp 351–356
36. Liu M, Yu H, Wang W (2009) FPAA based on integration of CMOS and nanojunction devices for neuromorphic applications. In: Cheng M (ed) Nano-Net. Springer, Berlin, pp 44–48
37. Liu W, Anguelov D, Erhan D, Szegedy C, Reed S, Fu C-Y, Berg AC (2016) SSD: single shot multibox detector. In: European conference on computer vision. Springer, Berlin, pp 21–37
38. Maguire LP, McGinnity TM, Glackin B, Ghani A, Belatreche A, Harkin J (2007) Challenges for large-scale implementations of spiking neural networks on FPGAs. Neurocomputing 71(1–3):13–29
39. Merolla P, Arthur J, Alvarez R, Bussat J, Boahen K (2014) A multicast tree router for multichip neuromorphic systems. IEEE Trans Circuits Syst I Regul Pap 61(3):820–833
40. Mortara A, Vittoz EA, Venier P (1995) A communication scheme for analog VLSI perceptive systems. IEEE J Solid-State Circuits 30(6):660–669
41. Nawrocki RA, Shaheen SE, Voyles RM (2011) A neuromorphic architecture from single transistor neurons with organic bistable devices for weights. In: The 2011 international joint conference on neural networks, July 2011, pp 450–456
42. Schemmel J, Grübl A, Hartmann S, Kononov A, Mayr C, Meier K, Millner S, Partzsch J, Schiefer S, Scholze S, Schüffny R, Schwartz M (2012) Live demonstration: a scaled-down version of the brainscales wafer-scale neuromorphic system. In: 2012 IEEE international symposium on circuits and systems, May 2012, p 702
43. Seo J, Brezzo B, Liu Y, Parker BD, Esser SK, Montoye RK, Rajendran B, Tierno JA, Chang L, Modha DS, Friedman DJ (2011) A 45nm CMOS neuromorphic chip with a scalable architecture for learning in networks of spiking neurons. In: 2011 IEEE custom integrated circuits conference (CICC), Sept 2011, pp 1–4
44. Song S, Miller KD, Abbott LF (2000) Competitive Hebbian learning through spike-timing-dependent synaptic plasticity. Nat Neurosci 3(9):919–926

45. Suri M, Bichler O, Querlioz D, Cueto O, Perniola L, Sousa V, Vuillaume D, Gamrat C, DeSalvo B (2011) Phase change memory as synapse for ultra-dense neuromorphic systems: application to complex visual pattern extraction. In: 2011 International electron devices meeting, Dec 2011, pp 4.4.1–4.4.4
46. Tamukoh H, Sekine M (2010) A dynamically reconfigurable platform for self-organizing neural network hardware. In: Wong KW, Mendis BSU, Bouzerdoum A (eds) Neural information processing. models and applications. Springer, Berlin, pp 439–446
47. Vincent AF, Larroque J, Locatelli N, Ben Romdhane N, Bichler O, Gamrat C, Zhao WS, Klein J, Galdin-Retailleau S, Querlioz D (2015) Spin-transfer torque magnetic memory as a stochastic memristive synapse for neuromorphic systems. IEEE Trans Biomed Circuits Syst 9(2):166–174
48. Vu TH, Murakami R, Okuyama Y, Abdallah AB (2018) Efficient optimization and hardware acceleration of CNNs towards the design of a scalable neuro inspired architecture in hardware. In: 2018 IEEE international conference on big data and smart computing (BigComp), Jan 2018, pp 326–332
49. Vu TH, Ikechukwu OM, Ben Abdallah A (2019) Fault-tolerant spike routing algorithm and architecture for three dimensional NoC-based neuromorphic systems. IEEE Access 7:90436–90452
50. Vu TH, Okuyama Y, Abdallah AB (2019) Comprehensive analytic performance assessment and k-means based multicast routing algorithm and architecture for 3d-NoC of spiking neurons. ACM J Emerg Technol Comput Syst 15(4):1–28
51. Yang Z, Murray A, Worgotter F, Cameron K, Boonsobhak V (2006) A neuromorphic depth-from-motion vision model with STDP adaptation. IEEE Trans Neural Netw 17(2):482–495
52. Yin S, Venkataramanaiah SK, Chen GK, Krishnamurthy R, Cao Y, Chakrabarti C, Seo J (2017) Algorithm and hardware design of discrete-time spiking neural networks based on back propagation with binary activations. CoRR, abs/1709.06206
53. Yu T, Cauwenberghs G (2009) Analog VLSI neuromorphic network with programmable membrane channel kinetics. In: 2009 IEEE international symposium on circuits and systems, May 2009, pp 349–352
54. Yu S, Wu Y, Wong H-SP (2011) Investigating the switching dynamics and multilevel capability of bipolar metal oxide resistive switching memory. Appl Phys Lett 98(10):103514

Chapter 2
Neuromorphic System Design Fundamentals

Abstract The neuromorphic computing paradigm has the potential to improve the efficiency of computational tasks. Unlike the typical artificial neural networks (ANNs), where neurons fire at each propagation cycle, the neurons in a neuromorphic neural networks model, named spiking neural networks (SNNs), fire only when their membrane potential reaches a certain threshold. Spiking neurons are only activated when sufficient signals are integrated from other neurons, which leads to sparse neural activities at the network level. Furthermore, their asynchronous event-driven operations, distributed memory, and massive parallelism significantly accelerate information processing and reduce energy consumption in many applications (i.e., pattern recognition, object detection, navigation, motor control, and so on). The key design challenges of neuromorphic systems include: how the organization of individual neurons, circuits, applications, and overall architectures enable energy-efficient computations, how information is represented, and how adaptation to local and evolutionary changes are facilitated. Moreover, a massively parallel neuromorphic architecture will require building small-sized neuro processing cores with low-power consumption, efficient neuron coding schemes, and a lightweight on-chip learning algorithm, which is also a challenges. This chapter covers fundamental design principles to build an efficient neuromorphic system in hardware.

2.1 Introduction

The term neuromorphic engineering is a concept developed by Carver Mead [23] in the late 1980s, describing the use of VLSI systems containing electronic analog circuits to mimic neuro-biological architectures present in the nervous system. In recent years, interest in neuromorphic system design has gradually increased due to better understandings of the brain, and the operation of neurons and several specialized structures, such as the retina. Current neuromorphic systems are hybrid analog-digital electronic systems fabricated using CMOS VLSI technology. Thus, the neuromorphic paradigm opens up computing prospects beyond traditional computer systems.

Spiking neural networks (SNNs) offer an efficient way of making inferences because computations are event-driven, and the neurons in the networks are sparsely activated. Moreover, SNNs can trade-off classification error rates against the number of available operations, whereas continuous-valued DNNs require a fixed number of functions to achieve their classification error rate. With an increase in error rate of a few percentage points, SNNs can achieve more than 2x reductions in operations when compared to the original ANNs (CNNs) for LeNet and MNIST, and BinaryNet CIFAR-10 [31]. This clearly shows the potential of SNNs, particularly when they are deployed on power-efficient embedded neuromorphic devices. Also, SNNs can generate results after the first output spike is produced, unlike ANNs where the output is available only after all layers have been completely processed [10]. Moreover, SNNs are naturally suited to process event-based sensor data. Still, even in classical frame-based machine vision applications, they are accurate, fast, and efficient, primarily when implemented on neuromorphic hardware.

Several neuromorphic processors have been designed for applications ranging from vision systems for object recognition in drones, simulation of some part of the brain, to running clustering algorithms that detect trade patterns. The operations of an SNN are generally grouped into learning and inference. Under these, a neuromorphic system is evaluated for efficiency while targeted towards maintaining low power, small footprint, and scalability. For a neuromorphic processor to efficiently carry out learning and inference, factors like learning rules and neuron models which significantly determine the performance of an SNN, have to be carefully considered.

Years of research in learning methods and inference in neuromorphic processors have narrowed it down to three approaches. The first involves conducting training and inference on high-end servers and then exporting the result. The second requires training on high-end servers and inference on an autonomous agent, while the third requires conducting both training and inference on an autonomous agent. However, the first approach requires a network connection that utilizes more resources and poses security challenges. The second approach presents the possibility of conserving energy, but with increasing demand for real-time adaptation in a dynamic environment from autonomous agents, the third approach is imperative.

2.1.1 Spiking Neural Networks

In recent years, neuroscience research has revealed a great deal about the structure and operation of individual neurons. Medical tools have also shown a great deal about how neural activity in different regions of the brain follow a sensory stimulus [12]. Moreover, the advances of software-based artificial intelligence (AI) have brought us to the edge of building brain-like functioning devices and systems that are not limited by the bottleneck of the conventional von Neumann computing architecture. The main difference between neuro-inspired (neuromorphic) systems and traditional information processing systems is their use of memory structures and organization [16]. While computing systems based on the von Neumann style have one or more central processing units physically separated from the main memory areas, both biological (spiking) and artificial neural network systems are characterized by distributed co-localized memory and computation.

The neuro-inspired technology based on SNNs (Spiking neural networks) witnesses increasing attention to better understand the brain and explore novel biologically inspired computation [44–46]. SNNs have been successfully applied to several applications, including visual recognition and classification tasks. Besides, implementations of neuromorphic hardware have enabled large-scale networks to run in real-time, which is a critical requirement for several applications.

SNNs attempt to mimic the information processing in the mammalian brain based on parallel arrays of neurons that communicate via spike events. Unlike the typical multi-layer perceptron networks, where neurons fire at each propagation cycle, the neurons in the SNN model fire only when their membrane potential reaches a certain threshold.

In SNN, a neuron generates spikes/pulses that can travel down nerve fibers if it receives enough stimuli from other neurons with the presence of external stimuli. These pulses (typically in the range of 1–2 ms) may vary in amplitude, shape, and duration, but they are treated as similar events. In these spiking models, information is encoded using various encoding schemes, such as coincidence coding, rate coding, or temporal coding. In an SNN, neurons communicate at junctions called synapses (chemical or electric). Chemical synapses, which form the majority of all synapses, communicate using chemical messengers. While in electrical synapses, ions flow directly between cells. Figure 2.1 illustrates two neurons communicating via a synapse.

Software simulation of SNN is a flexible method for investigating the behavior of neuronal systems. However, simulation of a deep SNN system requires analytical performance assessment and high-throughput, which obviously cannot be achieved with a software-based approach. An alternative approach is a hardware implementation, which can generate independent spikes accurately and simultaneously output spikes in real-time. Hardware implementations also have the advantage of computational speedup over software simulations and can take full advantage

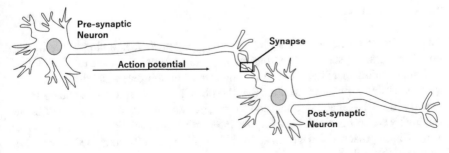

Fig. 2.1 Two neurons communicating via a synapse

of their inherent parallelism. Specialized hardware architectures with multiple neurocores could exploit the parallelism inherent within neural networks to provide high processing speeds with low power, making SNNs suitable for embedded neuromorphic devices and control applications. Some simulators also offer hardware acceleration that can speed up the performance [2].

A modern artificial neural network has many layers, leading to a "deep neural network (DNN)." In DNN, one neural network layer is often a 2D structure, especially in image recognition and classification networks; the resulting network is often a 3D structure. Therefore, mapping a 3D structure onto 2D circuits generally results in multiple long wires between layers or congestion.

2.1.2   Neural Coding Schemes

Spiking Neurons propagate signals rapidly over long distances by generating electrical pulses called action potentials that can travel down axons. Sensory neurons change their activities by firing sequences of action potentials, known as spike-train, in various temporal patterns, with external sensory stimuli, such as light, sound, etc.

In a small area of the cortex, there are thousands of spikes emitted in every millisecond. This raises the question of how the spikes can be encoded into information. Some other specialized neurons can communicate more information through so-called graded potentials. However, contrary to the action potential of spiking neurons, the signal in graded potential neurons decays much faster, necessitating high neuronal density and short inter-neuron distances. Graded potential neurons have the advantage of higher information rates capable of encoding more states than spiking neurons [32].

A spike train may contain information based on different coding schemes. For example, in motor neurons, the strength at which a muscle is contracted depends on the average number of spikes per unit time (firing rate). On the other hand, a complex temporal code is based on the precise timing of single spikes. This section describes the most common coding schemes used in spiking neuromorphic systems.

2.1.2.1   Rate Coding

The rate coding scheme, also known as frequency coding, is a simple traditional coding scheme. It states that as the intensity of a stimulus increases, the rate of action potentials increases. Since the sequence of action potentials generated by a given stimulus varies from trial to trial, the firing rates are used for decoding brain activity instead of specific spike sequences. Consequently, rate coding is inefficient but highly robust due to ISI (Inter-Spike-Interference)Inter-Spike-Interference noise [36]. This approach also neglects all the information possibly contained in the exact timing of the spikes.

In rate coding, precisely calculating the firing rate is essential. There are different rate averaging methods used in rate codings, such as an average over time (rate as a single-neuron spike count) or several repetitions:

Spike Count Rate   This method is determined by the average number of spikes in an interval time as shown in Eq. (2.1).

$$v_{sc} = \frac{n_{spike}}{\Delta t} \tag{2.1}$$

where n_{spike} is the spike number, Δt is the interval (time window). The length of Δt depends on the neural models used. This coding method has been successfully used for experiments involving sensory and motor system.

Spike Density Rate   In this coding method, the same stimulation sequence is repeated K times, then the number of spikes n_K is summed over all repetitions. The rate coding method is expressed in Eq. (2.2) as:

$$v_{sd} = \frac{n_K}{K \Delta t} \tag{2.2}$$

where Δt is the period of repetition. Although biological neurons do not use this method, it is a valuable method for evaluating neuron activity.

Population Activity Rate   Many neurons have the same characteristics and interact with the same stimuli. Population activity rate is proposed to measure the firing rate of a population of neurons as shown in Eq. (2.3):

$$v_{pa} = \frac{n_p}{N \Delta t} \tag{2.3}$$

where n_p is the total number of spikes generated by N neurons, Δt is the time window.

2.1.2.2   Temporal Coding

In temporal coding, the information about stimulus or action is contained in the relative timing of spikes, not just in the rate of those spikes. Temporal code is generally divided into the following coding schemes:

- *Time-to-first-spike:* The idea is to encode the latency information of the first spike event given a stream of spikes, where the precise timing of the spikes indicates the strength of the stimulation [13, 28, 34]. In general, the first spike is interpreted to carry a more powerful feature of the stimulus, and the following spikes are ignored [12, 26] (Fig. 2.2).
- *Inter-spike-interval:* The inter-spike-interval (ISI) refers to the internal time correlation between spikes rather than the absolute time with respect to stimulus

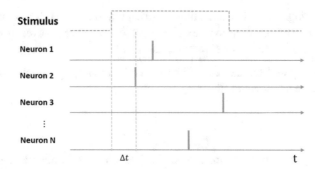

Fig. 2.2 Time to first spike

Fig. 2.3 Inter-spike-interval

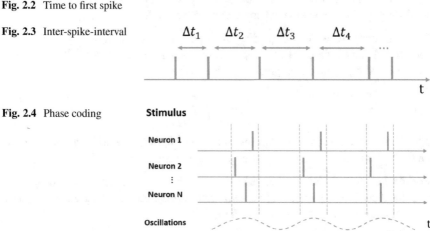

Fig. 2.4 Phase coding

onset [50, 51]. Also, the merit of the ISI approach compared to time-to-first-spike (TTFS), relies on the internal reference frames. Thus more messages can be conveyed during the same sampling period in this coding approach [1] (Fig. 2.3).

- *Phase coding:* The neural oscillations are intrinsically rhythmic and are prevalent in many brain areas with different frequency bands. When the activity of a cluster of neurons is periodical, the neurons can create rhythmic electrical patterns [3, 14]. The oscillations could serve as time inference signals, and the phase could be encoded into neuronal spike trains to convey relevant information (Fig. 2.4).
- *Rank order coding:* The idea is to omit the precise timing information by using only the order in which the spikes arrive [37, 38]. This coding strategy relies on the observation that the first-spike latency jitter between two trials is less than 1ms (median value) in all cases and thus is believed to be highly reliable [42]. An example is its robustness against image degradations like intensity and contrast change [37] (Fig. 2.5).
- *Correlation and synchrony:* Consider a pair or a group of neurons that are nearly synchronous. Such synchrony might imply some additional or specific information that is unable to be represented using the simple firing rate of neurons [12] (Fig. 2.6).

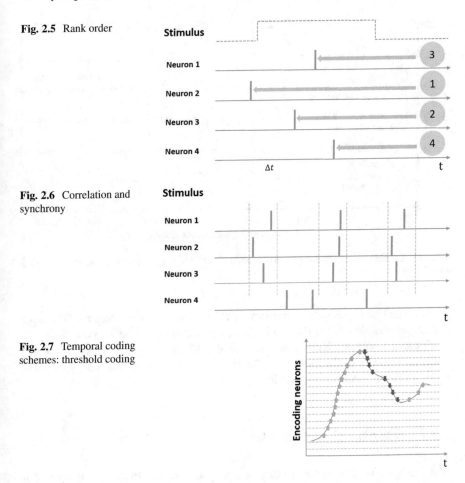

Fig. 2.5 Rank order

Fig. 2.6 Correlation and synchrony

Fig. 2.7 Temporal coding schemes: threshold coding

- *Threshold coding:* In general, when processing a continuous audio signal, an appropriate sample-rate is required for ideal preservation and perfect reconstruction of information. Based on the observation that the interval of sampling is the key to encode the stimuli, the sound intensity can be divided into a group of thresholds, and onset and offset neurons are used to encode the information each time the sound waves cross a threshold [26] (Figs. 2.7).

2.2 Spiking Neuron Models

Spiking neuron models are divided into different classes: the detailed mathematical models are biophysical neuron models that describe the membrane voltage as a function of the input current and the activation of ion channels. The simple models

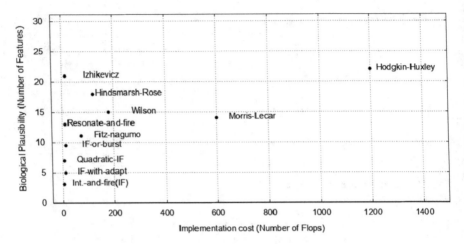

Fig. 2.8 A comparison of spiking neuron models in terms of implementation cost and biological plausibility

are neurons that express the membrane potential voltage as a function of the input current without describing the real biological processes of an action potential.

Figure 2.1 shows a biological neuron that consists of dendrites, axons, and cell body. A summary of the biological plausibility and computational cost of a pool of spiking neuron models evaluated in [18] is presented in Fig. 2.8.

2.2.1 Hodgkin-Huxley Model

The Hodgkin-Huxley neural model was proposed in the 1950s. As described in Fig. 2.9, the Hodkin-Huxley neuron models in a circuit, the cell membrane of a neuron as a capacitor that separates ionic charge from both sides of the membrane, the ion channels as conductances/resistors, and the membrane potential as battery. The calcium and sodium channels are voltage gated, and change connectivity based on the membrane potential, but the leak channels are not gated. This neuron model presents a mathematical description of the electric current through the membrane potential v giving the details of spike generation, as given in Eq. (2.4):

$$\frac{dv}{dt} = (\frac{1}{C})I - g_k n^4 (V - V_k) - g_{Na} m^3 h(V - V_{Na}) - g_L(V - V_L) \qquad (2.4)$$

where C is the capacitance of the circuit, I is the external current, conductances are potassium g_k, sodium g_{Na}, and leakage g_L. Gating parameters n, m, and h are determined by Eqs. (2.5)–(2.7), respectively

Fig. 2.9 The
Hodgkin-Huxley model: (**a**)
the schematic diagram
presents the membrane
potential, in which current
injection starts at $t = 5$ ms as
(**b**), while (**c**) and (**d**) show
the dependency of the gating
variables n, m and h on the
membrane potential

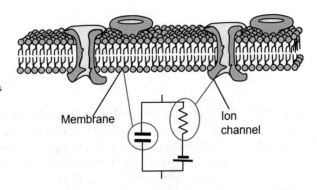

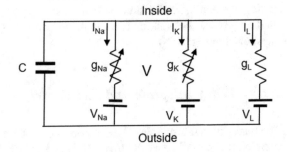

$$\frac{dn}{dt} = (n_\infty(v) - n)/\tau_n(v) \tag{2.5}$$

$$\frac{dm}{dt} = (m_\infty(v) - m)/\tau_m(v) \tag{2.6}$$

$$\frac{dh}{dt} = (h_\infty(v) - h)/\tau_h(v) \tag{2.7}$$

As shown in Fig. 2.8, the Hodgkin-Huxley model is the most biological plausible. However, its complexity with many features consumes a considerable amount of hardware resources. It, therefore, is costly for large-scale implementations.

2.2.2 Izhikevich Model

Compared to Hodgkin-Huxley, a less complex model was proposed by Izhikevich [18]. The following equations describe the model:

$$\frac{dv}{dt} = 0.04v^2 + 5v + 140 - u + I \tag{2.8}$$

$$\frac{du}{dt} = a(bv - u) \tag{2.9}$$

$$\begin{cases} v \leftarrow c \\ u \leftarrow u + d \end{cases} \quad if \ v \geq 30\,\text{mV} \tag{2.10}$$

where v is the membrane potential of the neuron, u is a membrane recovery variable, I is the neuron current, a, b, c, d are parameters of the models, in which the various values of these parameters result in different types of neuron characteristics. When membrane potential v exceeds the threshold (30 mV), the membrane potential v and recovery variable v are reset as Eq. (2.10). The Izhikevicz model is able to reproduce spiking and bursting characteristics of known cortical neurons. These spiking characteristics include the regular spiking (RS), intrinsically bursting (IB), chattering (CH), fast spiking (FS), thalamo-cortical (TC), resonator (RZ), and low-threshold spiking (LTS).

2.2.3 Leaky Integrate and Fire Model

The Leaky Integrate and Fire (LIF) model is one of the most commonly used models in SNN. It is described by the Eqs. (2.11) and (2.12).

$$\frac{dv}{dt} = I + a - bv \tag{2.11}$$

$$v \leftarrow c, \ if \ v \geq v_{th} \tag{2.12}$$

where v is the membrane potential of the neuron, I is the neuron current, a, b, and c are parameters of the model. When the membrane potential v exceeds a threshold v_{th}, it will be reset to c. The primary circuit presenting the LIF model is shown in Fig. 2.10. It consists of capacitor C and resistor R connected in parallel and driven by a current $I(t)$. In summary, among the existing spiking models, Hodgkin-Huxley, Izhikevich [18], and Leaky Integrate-and-Fire (LIF) [17] are often used. The Hodgkin-Huxley type is the best when measurable physiological parameters are highly considered. The model is based on many coefficients. This leads to challenges when implementing large SNNs because of the high hardware cost. In contrast, we can simulate hundreds of thousands of neurons when using LIF neural model; but, it is incapable of producing rich spiking patterns. Finally, Izhikevich exhibits a good compromise in terms of biophysical similarity and computational cost. It is close to the Hodgkin-Huxley model in biological plausibility while analogous to the LIF in computational complexity.

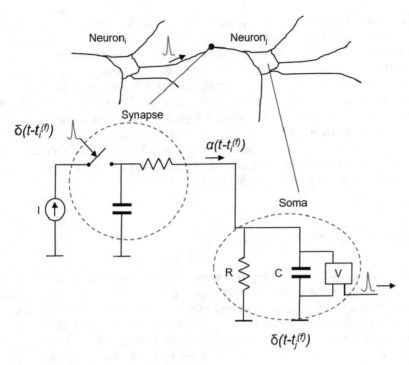

Fig. 2.10 Schematic diagram of the LIF model

2.3   Learning Algorithms

Learning in SNN is based on the modification of the strength of synaptic connections between neurons. Over the years, researchers have proposed and employed supervised and unsupervised learning rules for training neuromorphic systems to attain high performance.

2.3.1   Supervised Learning

Backpropagation (BP), a supervised learning method, is the most commonly used algorithm for neuromorphic programming. It can be employed in many neural network models such as feed-forward neural networks, recurrent neural networks, and convolution neural networks. The simple way to implement BP in hardware is off-chip. In this case, BP is performed on a traditional host machine. After that, pre-trained parameters are transferred or configured into the target neuromorphic chip. While this method is beneficial for precision software implementation, and requires lower hardware resources, it is not suitable for systems that have to re-train frequently. However, on-chip BP implementations have been used in many neuromorphic systems. Besides, variations of BP that are optimized or simplified for neuromorphic systems are also implemented. There are other on-chip learning implementations for convolution neural networks, Boltzmann machines, Restricted Boltzmann machines, and deep belief networks. The work in [11] successfully implemented spike-driven synaptic plasticity (SDSP) learning rule based on a so-called "teacher" signal. Unlike spike-timing-dependent plasticity (STDP), this learning rule induces an update each time a presynaptic spike occurs. On the other hand, backpropagation is also adopted for spiking systems [49]. Some authors first train an ANN with BP, then convert it into SNN by mapping real-value inputsactivation to average firing rates of Poisson spikes. This mechanism can be adopted to implement on spiking hardware as an off-chip learning method.

2.3.2 Unsupervised Learning

Compared to supervised learning, implementations of unsupervised learning are less popular. There have been some on-chip ones implemented in neuromorphic systems. Most of them were based on self-organizing maps or self-organizing learning rules.The spike-timing-dependent plasticity (STDP), which is a Hebbian-based learning, is a popular learning rule already implemented in several neuromorphic systems [16]. The operation of STDP depends on the firing time of presynaptic and postsynaptic neurons in which the synaptic weight will be increased when a spike arrives before a postsynaptic neuron "fires" and decreased when it arrives after a postsynaptic neuron fires. STDP is an unsupervised learning rule and is generally implemented on-chip.

2.3.2.1 Spike Timing Dependent Plasticity (STDP)

The STDP learning rule is originally expressed in [15], as:

$$\Delta w = \begin{cases} \Delta w^+ = A^+ e^{(\frac{-\Delta t}{\tau_+})}, & \text{if } \Delta t > 0 \\ \Delta w^- = -A^- e^{(\frac{\Delta t}{\tau_-})}, & \text{if } \Delta t \leq 0 \end{cases} \qquad (2.13)$$

Where Δw is the change in synaptic weight. If a presynaptic spike arrives the postsynaptic neuron within a time window τ_+ before the postsynaptic spike, the synaptic weight increases Δw^+, but if it arrives within a time window τ_-, after the postsynaptic spike, the synaptic weight decreases Δw^-. Δt is the time difference between the presynaptic and postsynaptic spike which is expressed as $\Delta t = t_{post} - t_{pre}$, while A^+ and A^- are potentiation and depression amplitude parameters respectively.

2.3.2.2 Spike Driven Synaptic Plasticity (SDSP)

Another SNN learning rule which has been employed in neuromorphic processors is spike driven synaptic plasticity (SDSP), which adapts synaptic weights at the arrival of presynaptic spikes [7]. At the arrival of a presynaptic spike, if the postsynaptic membrane potential exceeds the threshold, the synaptic weight is increased and decreased otherwise, ensuring that the calcium absorption in the postsynaptic site C(t) is preserved within a boundary when the presynaptic spike arrives. Its operation is described in [30] as:

$$
\begin{cases}
W = W + a, & \text{if } V_{\mathrm{mem}}(t) > V_{\mathrm{mth}} \text{ and } \theta_{up}^{l} < C(t) < \theta_{up}^{h} \\
W = W - b, & \text{if } V_{\mathrm{mem}}(t) \leq V_{\mathrm{mth}} \text{ and } \theta_{dn}^{l} < C(t) < \theta_{dn}^{h}
\end{cases}
\tag{2.14}
$$

where V_{mem} is the post synaptic membrane potential, V_{th} the threshold voltage, a and b the measure of potentiation and depression respectively. θ_{up}^{l}, θ_{up}^{h}, θ_{dn}^{l} and θ_{dn}^{h} are the boundaries for calcium absorption. In the absence of pre-synaptic spikes which determine potentiation and depression, the synaptic weights increase or decrease depending on its value which could be higher or lower than a set threshold θ_W. This operation is described in [7] as:

$$
\begin{cases}
\dfrac{dW(t)}{dt} = \alpha, & \text{if } W(t) > \theta_W \\[2mm]
\dfrac{dW(t)}{dt} = -\beta, & \text{if } W(t) \leq \theta_W.
\end{cases}
\tag{2.15}
$$

Other learning rules that have been implemented on neuromorphic processors include the BCM-like Local Correlation learning rule [22], Modified Ion Channel-based learning rule [35], and Stochastic gradient descent [29].

2.4 Synapse Memory

Unlike traditional computer systems, which have memory and processing different, SNN like biological neural networks combines both, having neurons as the processing elements and the synapses as memory. Efficiently replicating these synapses in neuromorphic processors has been challenging because a memory that behaves like synapses is required.

In traditional computer systems, memory speed is a bottleneck because processors have significantly improved over the years, surpassing memory speed and throughput, and this leaves processors idle while waiting for memory. However, SNN is different. These spiking neurons process with events, and as opposed to traditional processors that operate and communicate in megahertz and gigahertz ranges, they communicate in around 10 Hz. This difference in speed enables neuromorphic processors to use time multiplexing to combine many events into a single communication channel.

In SNN, Storing and reading synaptic weights constitutes the primary operation, and designing a large SNN with an enormous number of synapses will require large memory bandwidth. So while communication speed is not somewhat a challenge in neuromorphic systems, memory bandwidth needs to be overcome. In addressing this challenge, researchers have taken several approaches to realizes these synapses in CMOS by exploring various memory technologies. Some of these memory technologies include; static random-access memory (SRAM), a prominent memory technology in semiconductor design. In [33] a neuromorphic chip with 256 neurons that implements a transposable 8-transistor SRAM-based crossbar, which grants row, and column access, was employed.

Over the last few years, numerous efforts have been made to realize artificial synapses using post-CMOS devices, including resistive random-access memory (ReRAM), phase change memory devices, magnetoresistive random-access memory (MRAM), ferroelectric field-effect transistor (FeFET), and others. More recently, attempts have also been made to develop a non-CMOS neuron based on emerging devices. In this section, we present the major emerging memory technologies that can potentially replicate the behavior of synapses in neuromorphic systems. Each of these devices has its strengths and weaknesses. One type of device can be preferred over the others depending on the target application.

2.4.1 SRAM

A typical SRAM contains six transistors (6-T), and although having high leakage current and low density, SRAM offers multiple read and write, and its major advantages are speed and reliability when compared to other memory technologies. Another SRAM that has been utilized in neuromorphic systems is the eight transistor (8-T) SRAM [5]. The two more transistors on the 8-T SRAM adds access to the word and bit lines in transposed orientation to the typical 6-T design. To handle the general leakage power of the chip, the authors leveraged ultra-high-Vt devices which reduced it by 3, at the cost of the increased minimum operating voltage of the memory array (Fig. 2.11).

2.4.2 eDRAM

Another memory technology is embedded dynamic random-access memory (eDRAM), which was used in [8] to design a high-density 3D memory for a programmable digital neuromorphic architecture. eDRAM is a capacitor-based memory integrated on the same multi-chip module. Due to its simple conventional

Fig. 2.11 SRAM

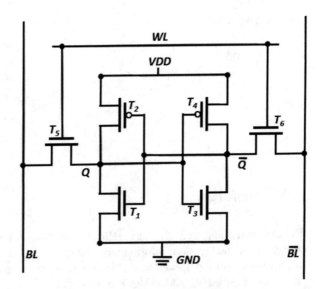

Fig. 2.12 EDRAM

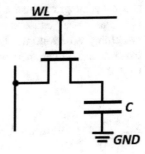

one 1-transistor 1-capacitor design, it is inclined to having less area cost when compared to SRAM. However, with the gradual leakage of its storage charge during operation, its retention period is low. Also, its design is not easily compatible with CMOS in [43], a spin-transfer torque ram (STT-RAM) was proposed as a stochastic memristive synapse for neuromorphic systems. An STT-RAM is a magnetic RAM that uses magnetic tunneling junction (MTJ) in its cells, and its simple design has a small area when compared to SRAM and eDRAM. By adopting a schema proposed in [21], the authors organized the STT-RAM MJT as a crossbar connecting input and output neurons. However, the magnetization of the STT-RAM makes the writing process slow and consumes more energy (Fig. 2.12).

Fig. 2.13 MEMRISTOR

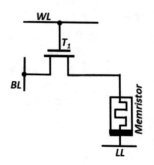

2.4.3 Memristor

Resistive random-access memory (RRAM or ReRAM) relies on the resistance change of its cells to store information, and it has similar architecture to eDRAM. In [20] the authors proposed an FPGA-based hardware emulator for a neuromorphic chip with an RRAM-based crossbar. RRAM offers low area, low power, and easy integration on CMOS. However, it suffers from stuck at faults (SAF) [48] which may cause short-circuiting and lead to increased power and dynamic switching variation, which leads to a significant variation in the resistance of the memory. Other memory technologies include Phase Change Memory (PCM) (Fig. 2.13).

2.5 Inter-Neuron Communication Schemes

Communication architectures for multicore spiking neuromorphic systems are responsible for delivering spikes between neuro-cores/tiles. They can be categories as intra-chip and inter-chip. For inter-chip, address event presentation (AER) is commonly employed [4, 6, 24]. In AER, each neuron has a unique address. Whenever a neuron generates a spike, its address is sent to postsynaptic neurons by a high-speed digital bus. AER is suitable for SNN implementations since it only needs to be active whenever neurons fire. To scale up the system, a hierarchical AER as a tree structure was implemented in [27].

On the other hand, network-on-chip (NoC) is commonly implemented for on-chip communication. In the early stage of the implementation, buses are employed in some systems [25]. However, works in [39, 40] evaluated and compared four architectures: bus, tree, point to point, and mesh. The results show that mesh with multicast offers the highest performance for SNN implementations. Furthermore, AER also is used for on-chip communication [9, 41].

2.5.1 AER—Address Event Representation

A neuromorphic system consists of many neurons, axons, and synapses integrated into a 3D or 2D silicon substrate. Unlike standard digital logic, where the output of a gate is generally connected to the input of three to four other gates, a neuron is typically connected to thousands of other neurons. From another hand, the spiking rate is shallow (tens of Hz) compared to the speed of digital electronics, which is in the range of GHz. Time-multiplexed communication protocol is often used to solve the speed difference between electronic and ionic transmissions.

The circuits used to multiplex communication for a cluster of neurons into a single communication channel is known as AER [19]. AER is a neuromorphic inter-chip communication protocol that allows for massive real-time connectivity between many neurons located on different chips. Figure 2.14 shows the AER protocol. Address-event representation (AER) is a communication protocol initially proposed to communicate sparse neural events between neuromorphic chips.

Initially, inter-chip communication networks provided only simple unidirectional, point-to-point connectivity between arrays of neuromorphic on two neuromorphic chips. These communication schemes map spikes from output nodes in the sending chip to any appropriate input nodes in the receiving chip. The mapping occurs asynchronously and provides random access to the receiver nodes. The spikes are actually represented as addresses. An address-encoder at the output node generates a unique binary address that identifies that node (neuron).

The output addresses are transmitted over a shared bus to the receiving chip, where an address decoder selects the appropriate receiver node (input) and activates it. Two versions of this random-access scheme have been proposed, a hard-wired version, and an arbitered version Each spike is represented by its location (explicitly encoded as an address) and the time it occurs (implicitly encoded).

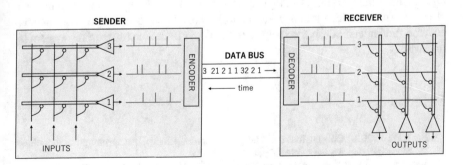

Fig. 2.14 AER protocol

2.6   Neuromorphic Spike Routing

A scalable inter-neuron communication architecture is required to design in silicon a neuromorphic system capable of housing even a fraction of the number of neurons in the brain. Furthermore, since the timing of spikes is used to encode information in SNN, such inter-neuron communication architecture should not violate the timing of spikes, as this will affect the performance of the SNN.

Various communication interconnects could be employed when designing inter-neuron communication architectures, including shared bus and packet-switched network on chip (NoC). However, a shared bus is a poor choice when implementing a large-scale SNN since it suffers adversely from an increased number of nodes. The nonlinear increase in neural connectivity will be too much for such an interconnect to handle. An interconnect that has been considered as a potential solution is the 2D packet-switched NoC (2D-NoC). However, with further scaling, 2D-NoC may experience communication challenges that affect power and performance, especially in large-scale SNN chips. 3D packet-switched NoC (3D-NoC), on the other hand, enables scaling and parallelism in the third dimension by combining NoC and 3D ICs (3D-ICs) [4, 47]. With the help of its short through-silicon vias (TSVs) that enable communication between layers, it can reduce communication costs. These merits of 3D-NoC make it suitable for large-scale SNN applications. Moreover, the brain is biologically organized in a 3D structure; therefore, by adopting the 3D interconnect, neuromorphic systems can inherit the shape and the interconnects of a biological brain.

2.7   Chapter Summary

In adopting the structure and computational principle of the brain, coupled with the benefits that hardware implementation provides, neuromorphic computing can provide a solution beyond the limitations faced by traditional computers. This structure and computational principle also enable it to perform real-time cognition tasks, which the traditional computer is not good at. However, in designing an efficient neuromorphic system, several hurdles need to be surmounted. This chapter covered fundamental design principles to build an efficient neuromorphic system in hardware.

References

1. Bai K, Yi Y (2019) Opening the "black box" of silicon chip design in neuromorphic computing. In: Bio-inspired technology. IntechOpen
2. Balaji A, Adiraju P, Kashyap HJ, Das A, Krichmar JL, Dutt ND, Catthoor F (2020) PyCARL: a PyNN interface for hardware-software co-simulation of spiking neural network. Preprint, arXiv:2003.09696

3. Başar E (2013) Brain oscillations in neuropsychiatric disease. Dialogues Clin Neurosci 15(3):291
4. Ben Abdallah A, Dang KN (2021) Toward robust cognitive 3d brain-inspired cross-paradigm system. Front Neurosci 15:795
5. Bhaskar A (2017) Design and analysis of low power SRAM cells. In: 2017 Innovations in power and advanced computing technologies (i-PACT). IEEE, Piscataway, pp 1–5
6. Boahen KA (1998) Communicating neuronal ensembles between neuromorphic chips. In: Neuromorphic systems engineering. Springer, Berlin, pp 229–259
7. Brader JM, Senn W, Fusi S (2007) Learning real-world stimuli in a neural network with spike-driven synaptic dynamics. Neural Comput 19(11):2881–2912
8. Chang M, Rosenfeld P, Lu S, Jacob B (2013) Technology comparison for large last-level caches (L3Cs): low-leakage SRAM, low write-energy STT-RAM, and refresh-optimized eDRAM. In: 2013 IEEE 19th international symposium on high performance computer architecture (HPCA), Feb 2013, pp 143–154
9. Deiss SR, Douglas RJ, Whatley AM, Maass W (1999) A pulse-coded communications infrastructure for neuromorphic systems. In: Pulsed neural networks, pp 157–178
10. Diehl PU, Neil D, Binas J, Cook M, Liu S, Pfeiffer M (2015) Fast-classifying, high-accuracy spiking deep networks through weight and threshold balancing. In: 2015 International joint conference on neural networks (IJCNN), July 2015, pp 1–8
11. Frenkel C, Legat J, Bol D (2019) Morphic: a 65-nm 738k-synapse/mm^2 quad-core binary-weight digital neuromorphic processor with stochastic spike-driven online learning. IEEE Trans Biomed Circuits Syst 13(5):999–1010
12. Gerstner W, Kistler WM, Naud R, Paninski L (2014) Neuronal dynamics: from single neurons to networks and models of cognition. Cambridge University Press, Cambridge
13. Göltz J, Baumbach A, Billaudelle S, Breitwieser O, Dold D, Kriener L, Kungl AF, Senn W, Schemmel J, Meier K et al (2019) Fast and deep neuromorphic learning with time-to-first-spike coding. Preprint, arXiv:1912.11443
14. Hakim N, Vogel EK (2018) Phase-coding memories in mind. PLoS Biol 16(8):e3000012
15. Iannella N, Launey T, Tanaka S (2010) Spike timing-dependent plasticity as the origin of the formation of clustered synaptic efficacy engrams. Front Comput Neuros 4:21
16. Ikechukwu OM, Dang KN, Abdallah AB (2021) On the design of a fault-tolerant scalable three dimensional NoC-based digital neuromorphic system with on-chip learning. IEEE Access 9:64331–64345
17. Indiveri G, Chicca E, Douglas R (2006) A VLSI array of low-power spiking neurons and bistable synapses with spike-timing dependent plasticity. IEEE Trans Neural Netw 17(1):211–221
18. Izhikevich E (2003) Simple model of spiking neurons. IEEE Trans Neural Netw 14(6):1569–1572
19. Lazzaro J, Wawrzynek J, Mahowald M, Sivilotti M, Gillespie D (1993) Silicon auditory processors as computer peripherals. IEEE Trans Neural Netw 4(3):523–528
20. Luo T, Wang X, Qu C, Lee MKF, Tang WT, Wong W-F, Goh RSM (2018) An FPGA-based hardware emulator for neuromorphic chip with RRAM. IEEE Trans Comput Aided Des Integr Circuits Syst 39(2):438–450
21. Majumder T, Suri M, Shekhar V (2015) NoC router using STT-MRAM based hybrid buffers with error correction and limited flit retransmission. In: 2015 IEEE international symposium on circuits and systems (ISCAS). IEEE, Piscataway, pp 2305–2308
22. Mayr CG, Partzsch J (2010) Rate and pulse based plasticity governed by local synaptic state variables. Front Synaptic Neurosci 2:33
23. Mead C (1990) Neuromorphic electronic systems. Proc IEEE 78(10):1629–1636
24. Merolla PA, Arthur JV, Alvarez-Icaza R, Cassidy AS, Sawada J, Akopyan F, Jackson BL, Imam N, Guo C, Nakamura Y et al (2014) A million spiking-neuron integrated circuit with a scalable communication network and interface. Science 345(6197):668–673
25. Mortara A, Vittoz EA, Venier P (1995) A communication scheme for analog VLSI perceptive systems. IEEE J Solid-State Circuits 30(6):660–669

26. Pan Z, Wu J, Zhang M, Li H, Chua Y (2019) Neural population coding for effective temporal classification. In: 2019 International joint conference on neural networks (IJCNN). IEEE, Piscataway, pp 1–8
27. Park J, Yu T, Joshi S, Maier C, Cauwenberghs G (2016) Hierarchical address event routing for reconfigurable large-scale neuromorphic systems. IEEE Trans Neural Netw Learn Syst 28(10):2408–2422
28. Park S, Kim S, Na B, Yoon S (2020) T2fsnn: deep spiking neural networks with time-to-first-spike coding. Preprint, arXiv:2003.11741
29. Ponulak F, Kasiński A (2010) Supervised learning in spiking neural networks with resume: sequence learning, classification, and spike shifting. Neural Comput 22(2):467–510
30. Rahimi Azghadi M, Iannella N, Al-Sarawi SF, Indiveri G, Abbott D (2014) Spike-based synaptic plasticity in silicon: Design, implementation, application, and challenges. Proc IEEE 102(5):717–737
31. Rueckauer B, Lungu I-A, Hu Y, Pfeiffer M, Liu S-C (2017) Conversion of continuous-valued deep networks to efficient event-driven networks for image classification. Front Neurosci 11:682
32. Sengupta B, Laughlin SB, Niven JE (2014) Consequences of converting graded to action potentials upon neural information coding and energy efficiency. PLoS Comput Biol 10(1):1–18
33. Seo J-s, Brezzo B, Liu Y, Parker BD, Esser SK, Montoye RK, Rajendran B, Tierno JA, Chang L, Modha DS et al (2011) A 45nm CMOS neuromorphic chip with a scalable architecture for learning in networks of spiking neurons. In: 2011 IEEE custom integrated circuits conference (CICC). IEEE, Piscataway, pp 1–4
34. Shoushun C, Bermak A (2005) A low power CMOS imager based on time-to-first-spike encoding and fair AER. In: 2005 IEEE international symposium on circuits and systems. IEEE, Piscataway, pp 5306–5309
35. Sjostrom PJ, Rancz EA, Roth A, Hausser M (2008) Dendritic excitability and synaptic plasticity. Physiol Rev 88(2):769–840
36. Stein RB, Gossen ER, Jones KE (2005) Neuronal variability: noise or part of the signal? Nat Rev Neurosci 6(5):389–397
37. Thorpe S, Gautrais J (1998) Rank order coding. In: Computational neuroscience. Springer, Berlin, pp 113–118
38. Thorpe S, Delorme A, Van Rullen R (2001) Spike-based strategies for rapid processing. Neural Netw 14(6–7):715–725
39. Vainbrand D, Ginosar R (2010) Comparing NoC architectures for neural networks. In: 2010 IEEE 26-th convention of electrical and electronics engineers in Israel. IEEE, Piscataway, pp 000660–000664
40. Vainbrand D, Ginosar R (2010) Network-on-chip architectures for neural networks. In: 2010 Fourth ACM/IEEE international symposium on networks-on-chip. IEEE, Piscataway, pp 135–144
41. van Schaik A, Liu S-C (2005) AER EAR: a matched silicon cochlea pair with address event representation interface. In: 2005 IEEE international symposium on circuits and systems. IEEE, Piscataway, pp 4213–4216
42. VanRullen R, Guyonneau R, Thorpe SJ (2005) Spike times make sense. Trends Neurosci 28(1):1–4
43. Vincent AF, Larroque J, Locatelli N, Romdhane NB, Bichler O, Gamrat C, Zhao WS, Klein J-O, Galdin-Retailleau S, Querlioz D (2015) Spin-transfer torque magnetic memory as a stochastic memristive synapse for neuromorphic systems. IEEE Trans Biomed Circuits Syst 9(2):166–174
44. Vu TH, Ikechukwu OM, Abdallah AB (2019) Fault-tolerant spike routing algorithm and architecture for three dimensional NoC-based neuromorphic systems. IEEE Access 7:90436–90452

45. Vu TH, Murakami Y, Abdallah AB (2019) Graceful fault-tolerant on-chip spike routing algorithm for mesh-based spiking neural networks. In: 2019 2nd International conference on intelligent autonomous systems (ICoIAS), Singapore, Feb 2019
46. Vu TH, Murakami Y, Abdallah AB (2019) A low-latency tree-based multicast spike routing for scalable multicore neuromorphic chips. In: ACM 5th international conference of computing for engineering and sciences, Hammamet, Tunisia, July 2019
47. Vu TH, Okuyama Y, Abdallah AB (2019) Comprehensive analytic performance assessment and k-means based multicast routing algorithm and architecture for 3d-NoC of spiking neurons. ACM J Emerg Technol Comput Syst 15(4):1–28
48. Xia L, Huangfu W, Tang T, Yin X, Chakrabarty K, Xie Y, Wang Y, Yang H (2017) Stuck-at fault tolerance in RRAM computing systems. IEEE J Emerg Sel Top Circuits Syst 8(1):102–115
49. Yin S, Venkataramanaiah S, Chen G, Krishnamurthy R, Cao Y, Chakrabarti C, Sun Seo J (2018) Algorithm and hardware design of discrete-time spiking neural networks based on back propagation with binary activations. In: 2017 IEEE biomedical circuits and systems conference, BioCAS 2017 - Proceedings, Jan, vol 2018. Institute of Electrical and Electronics Engineers, Piscataway, pp 1–4
50. Zhao, C, Wysocki BT, Thiem CD, McDonald NR, Li J, Liu L, Yi Y (2016) Energy efficient spiking temporal encoder design for neuromorphic computing systems. IEEE Trans Multi-Scale Comput Syst 2(4):265–276
51. Zhao C, Yi Y, Li J, Fu X, Liu L (2017) Interspike-interval-based analog spike-time-dependent encoder for neuromorphic processors. IEEE Trans Very Large Scale Integr Syst 25(8):2193–2205

Chapter 3
Learning in Neuromorphic Systems

Abstract The human brain is regarded as a power-efficient learning machine capable of carrying out complex computations while using only little resources. A sophisticated property that makes energy-efficient computation possible is the distinct sparse communication among many spiking neurons. The primary goal of neuromorphic hardware is to emulate brain-like neural networks to solve real-world problems. However, training on neuromorphic systems is challenging due to the required non-local computations of gradient-based learning algorithms. In Spiking neural networks, there are two fundamental modes: inference and learning. The learning phase, which minimizes a particular cost or loss function, is a complex process of acquiring the parameters to output the correct inference results. On the other hand, the inference computes the output values based on the given input and the network parameters. This chapter presents how learning in neuromorphic computing systems is conducted.

3.1 Learning Methods

Spiking neural network (SNN)has gradually gained awareness because of its ability to process and communicate sparse binary signals (spikes) in a highly parallel and event-driven manner analogous to the biological brain [44, 46, 47]. However, simulating large-scale SNN in software is slow and does not fully harness the energy efficiency of SNN. As an alternative, scalable multicore spike-based neuromorphic architectures that can support a massive number of neurons and synapses and leverage the spike sparsity available in SNN to deliver rapid parallel processing with low power are being proposed. However, realizing such a neuromorphic architecture requires building small-sized spiking neuro-cores with low-power consumption, efficient neural coding schemes, and learning methods [32].

The learning phase, which minimizes a particular cost (loss) function, is a complex process of acquiring the parameters to output the correct inference results. The cost function optimization is performed with a gradient-descent-based optimization or other classical optimization methods, such as a genetic algorithm. In gradient-decent-based optimization, the basic idea is to find out the gradients of

© The Author(s), under exclusive license to Springer Nature Switzerland AG 2022
A. Ben Abdallah, K. N. Dang, *Neuromorphic Computing Principles and Organization*, https://doi.org/10.1007/978-3-030-92525-3_3

the cost function concerning each learning parameter. There are various proposed training/learning algorithms for SNNs, such as supervised backpropagation through time, unsupervised STDP learning, and ANN to SNN conversion. In the following, we will provide description of common training rules and methods in neuromorphic systems.

3.2   Conversion from ANN to SNN

The goal of the ANN to SNN conversionconversion approach is to leverage the state-of-the-art ANN training algorithms so that the converted SNN version can reach the competitive classification performance of its off-line trained ANN.

There are many challenges in converting the network from ANN to SNN [6, 43]. Firstly, negative output values, which come from the preprocessing procedure, sigmoid function, or the calculation based on weights and biases, which might be both negative values, are difficult to represent in SNN accurately. Considering inhibitory neurons for representing negative values leads to much more sophisticated network architecture. Thus more hardware resources and computation complexity are needed. Secondly, because biases in each layer could have negative values, they are also difficult to represent in SNN. Moreover, implementing max-pooling in SNN is another critical issue we need to take into account. In conventional ANN, the max-pooling operation is executed between two adjacent layers. However, if we want to perform such a function in spiking neuron networks, a two-layer neural network and more neurons are also required, and the accuracy loss is expected to happen.

To cope with these challenges, the following changes are made to meet the requirements of SNN. First, using the abs() function and rectified linear unit instead of the sigmoid function tanh(), we avoid negative output when converting a CNN to SNN. For biases in each layer, they are then set to zero. The spatial max-pooling can be replaced by spatial linear sub-sampling, which is easily converted to a spike domain. However, the performance loss is still unavoidable when the ReLUs in the ANN have been replaced by IF neurons in SNN. To address the problem, Diehl et al. [7] proposed a weight normalization method to achieve a near-lossless accuracy. The first approach, model-based normalization, requires only the information of the network weights, while the data-based normalization approach scales the weights according to the actual activation of the network in response to data.

3.2.1   Converted SNNs

Spiking Deep Belief Networks (DBNs) Pérez-Carrasco et al. [28] began by converting the conventional ConvNet into an event-driven architecture, in which the frame-driven system neurons are mapped into event-driven representation. Cao et al. [6]

proposed a new approach for converting a CNN to SNN architecture that is suitable for mapping to spike-based neuromorphic hardware by raising and tackling several key issues during conversions, such as negative activations and biases and non-linearity due to max-pooling operation. This work is followed by Diehl et al. [7], where a method of weight normalization is introduced to reduce the latency with high accuracy. Hunsberger and Eliasmith [13] adopted AlexNet by training with noise on output neurons, yielding a more robust model against spiking variability. A modified "soft" LIF function is also proposed, allowing more neuron types to be utilized. Zambrano and Bohte [53] constructed an adaptive spiking neural network, where the threshold is dynamically adjusted, and fewer spikes are needed to encode information. While many research works have contributed to such conversion methods, the choice of neural coding schemes attracts insufficient attention. Wu et al. [49] proposed a novel spike-based learning rule. The proposed approach is hardware-friendly due to the requirement of less computation and memory while showing competitive performance on MNIST datasets. Rueckauer et al. [33] and Zhang et al. [54] applied temporal coding schemes to converted SNN, whereby the spike redundancy and memory cost are significantly reduced. Considering that most of these models use only a few convolution layers, their performance on rather complicated datasets like CIFAR is limited due to their shallow architectures. Rueckauer et al. [34] successfully implemented the conversion on Inception-V3 with 42 layers (7 convolution layers), which demonstrates a 74.60% accuracy on the ImageNet dataset. They also proposed a Spiking CNN with four convolution levels, achieving a 90.85% accuracy on the CIFAR-10 dataset. Sengupta et al. [38] presented two deep spiking neuron networks based on VGG-16 [41] and Residual network architecture. The residual network architecture was also employed in the work of Hu et al. [12]. Motivated by the "hard reset" spiking neuron model that leads to performance deterioration due to information loss during the conversion, Han et al. [11] introduced Residual Membrane Potential (RMP) spiking neuron, which targets the spike rate vanishing issue in SNNs. They implemented the conversion of both VGG and Residual network architecture with high accuracy on CIFAR and ImageNet datasets while having low conversion loss.

3.2.2 Challenges of ANN Conversion

There are several challenges faced when converting ANN into an SNN as in Fig. 3.1. First, there are many connection weights and biases between two adjacent layers in a traditional convolution neural network, which can be either positive or negative values. When the inputs are transmitted between neurons, the weights are applied to the inputs and passed into an activation function and bias. The resulting activation could be positive or negative through activation functions like *sigmoid* and *tanh*. While it is not a big issue in ANN, the firing rates in SNN should always be positive, which requires the designer to tackle negative values during

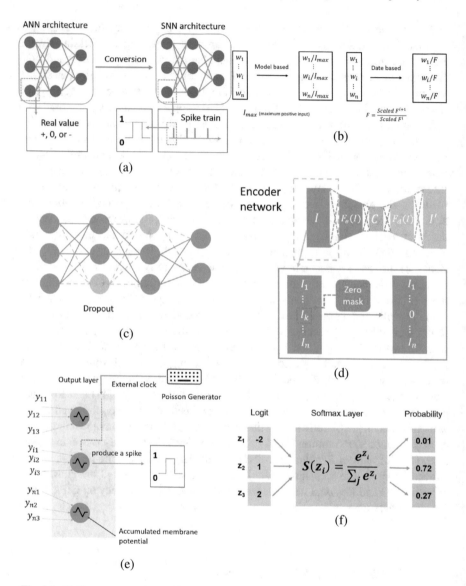

Fig. 3.1 Challenges, strategies and improvements in conversion from ANNs to SNNs

conversion. One possible solution is to treat positive values as excitatory synaptic input while producing inhibitory synaptic inputs for negative signals [28]. However, two spiking neurons are needed to represent each input value, which requires more interconnections in the spiking neuron network and thus a much more complicated architecture. Rectified linear-unit (ReLU) activation function [23] is an efficient tool to circumvent this problem. Because $ReLU(x) = max(x, 0)$ is always mapping a

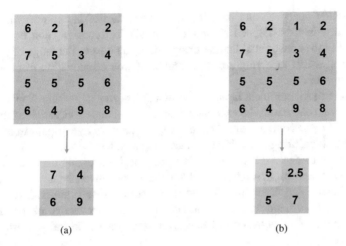

Fig. 3.2 Max-pooling and average-pooling operation. (**a**) Max-pooling. (**b**) Average-pooling

negative activation to zero, it can be considered as an ideal approximation of an integrate-and-fire neuron without a refractory period. Avoiding negative inputs not only relaxes the translation from activation into firing rates, but it also contributes to faster convergence compared with equivalent networks with tanh units [17].

Second, negative biases are difficult to represent in spiking neural networks. However, setting all biases to zero can reduce the inconvenience [6]. Besides, consider a max-pooling operation that has been widely used in conventional neural networks. To downsample the input feature maps, this discretization process reduces the number of parameters and computational cost by applying a max filter [35]. Since the neuron activation in SNNs is encoded in binary representation [38] and the operation is non-linear, a loss is unavoidable. The pooling operation between two network layers, along with strong lateral inhibition [52], causes extra complexity as well. Moreover, using max-pooling does not reflect the actual maximum firing rate due to the Winner-Take-All strategy. Even though Max-pooling holds an advantage of better translation feature invariant compared to average-pooling operation (See Fig. 3.2), average-pooling operation is a better option that enables a linear function to be implemented in SNNs [6].

Masquelier and Thorpe [19] approach this problem using the time-to-first-spike learning rule, in which the earliest spikes are considered to carry most information of the input patterns. In the first architecture [27], the temporal "time-to-first-spike" encoding is used to select the first neuron that responds; thus, this neuron is considered to have the most robust response to the stimulus. However, these works are typically based on a temporal coding scheme, which is not ideal for the ANN-SNN conversion process.

Rueckauer et al. [34] proposed a gating-function for spiking max-pooling, which allows only the spikes from the neuron with the highest firing rate by estimating the presynaptic firing rates. The proposed response mechanism allows a nonlinear pooling operation and introducing a better start of the conversion process. Additionally, in time-stepped simulations of SNNs, approximation errors might occur due to the constraints that the firing rate is mapped to the range of $[0, r_{max}]$, which implies that receiving a perfect representation of activation from ANNs to SNNs is non-trivial.

Consider that insufficient inputs arrive at a unit during a tiny time step; a relevant high threshold can hardly be exceeded, leading to an underestimation of the actual firing rate. On the contrary, over-activation spike trains or high input weights would give rise to high firing rates due to a lack of integrated evidence [7, 34, 38]. One possible way to deal with this issue is to rescale the weights using *model-based normalization* approach by adjusting all the parameters given the maximum possible positive inputs. An alternative method is to use *data-based normalization*, which rescales all the synaptic weights corresponding to a successive neuron layer by the maximum neural activation of that layer when the entire training procedure is completed [7].

Sengupta et al. [38] extended the proposed weight-normalization scheme by taking into account the actual operation of the SNN during the conversion process, whereby the temporal delay of the neuron is decreased, and an appropriate firing threshold is ensured. Rueckauer et al. [34] proposed to preserve the encoded information of biases jointly scaled with input weights and a *max-norm* mechanism to detect and discard outlier activations, achieving a more robust network. When strong normalization is further combined with batch-normalization used to remove covariance shift from internal activation of the network during training [15], the whole process achieves more significant speedup while reducing the number of outliers in each layer. These works studied the balance of firing rates, spiking threshold, and input weights, with an ideal tradeoff between accuracy and latency.

The work by Neil et al. [25] focused on a more efficient algorithm that contributes to improving computation efficiency. By making good use of sparse coding and L2-norm as a cost function on activation, lower-compute spiking neurons with fewer spikes are realized. Therefore the overall firing rates are reduced. Another strategy is to accelerate the accurate classification task, either by utilizing *Dropout* [42] with its corresponding learning schedule, or by leveraging a trained *Stacked Auto-Encoder* with a zero-masking filter. It is expected that fewer input spikes are needed for quick output. The softmax function in output layer is used to normalize the input values into a valid probability distribution that sum to one [21] (See Fig. 3.3). As discussed in [34], without the Softmax layer, pure negative inputs arriving at the final layers will not produce any spike. Thus the prediction fails. One version of a spiking softmax layer is a stochastic winner-take-all (WTA) mechanism [26] with an external Poisson generator, whereby the winning neuron is selected according to its membrane potential, and the WTA-circuit allows it to fire at that time step. Furthermore, such a mechanism can be simplified; that is, given the membrane potentials, the classification can directly be inferred based on the computed rate parameters at softmax layer [34].

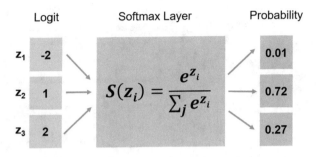

Fig. 3.3 Softmax function

3.3 Supervised Learning

The feed-forward spiking neural network contains connections of spiking neurons between layers with multiple delayed synaptic terminals [1, 14, 24, 45]. Consider a connection between neuron H_3 of hidden layer I and neuron O_2 of output layer J in Fig. 3.4b. Each pre-synaptic terminal corresponds to a sub-connection associated with different decay and synaptic efficacy, respectively. A spike-response function $\varepsilon(t)$, is used to describe a standard post-synaptic potential (PSP):

$$\varepsilon(t) = \frac{t}{\tau}e^{1-t/\tau} \tag{3.1}$$

where τ is defined as the membrane decay time constant of a neuron. When taking into account multiple synapses per connection, the post-synaptic input x_j of neuron j receiving input from neuron i can then be described as the weighted sum of the pre-synaptic input:

$$x_j(t) = \sum_i \sum_k w_{ij}^k \varepsilon_{ij}^k(t - t_i - d_{ij}^k) \tag{3.2}$$

where i belongs to the set of all input neurons to neuron j, and t_i is the time of the input spike rising from neuron i [2, 20]. Once the combinations of the internal state variable crosses the activation threshold θ, the output neuron produces a spike at time t_j.

3.3.1 Tempotron

Tempotron is a classic model of supervised learning for classification tasks, which uses a leaky integrate-and-fire neuron driven by synaptic afferents [10]. The subthreshold membrane voltage is a weighted sum of postsynaptic potentials contributed by all incoming spikes:

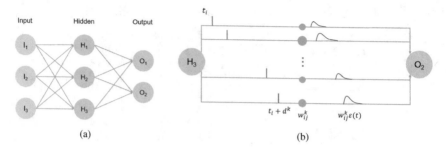

Fig. 3.4 Network architecture and connectivity of a spiking neural network. (**a**) Feedforward spiking neural network. (**b**) Connection consisting of multiple delayed synaptic terminals

$$V(t) = \sum_i w_i \sum_{t_i} V_0 (e^{-\frac{t-t_i}{\tau}} - e^{-\frac{t-t_i}{\tau_s}}) + V_{rest} \qquad (3.3)$$

where w_i is the synaptic weight between neuron i and one of its postsynaptic neurons, and a postsynaptic potential is induced by an input spike from neuron i at time t_i. The parameters τ and τ_s denote the decay time constants of membrane integration and synaptic currents, respectively, and are used to describe the form of the postsynaptic potentials, where the maximum value of PSP is normalized to 1 with a factor V_0. For the *Tempotron* learning rule, each synaptic efficacy w_i follows the gradient descent updating mechanism:

$$\Delta w_i = \lambda \sum_{t_i < t_{max}} (e^{-\frac{t_{max}-t_i}{\tau}} - e^{-\frac{t_{max}-t_i}{\tau_s}}) \qquad (3.4)$$

where t_{max} is the time when the postsynaptic potential reaches its maximal value without the neuron firing. Based on the learning rule, when the erroneous output spike occurs, the corresponding synaptic weight should be decreased for less contribution on the wrong pattern, and when neuron should fire but does not, the update scheme will increase weight w_i in the iteration.

3.3.2 ReSuMe

It was discussed in [8] that the *Tempotron* learning rule can be considered as a particular case of the remote supervised method (*ReSuMe*) [30]. Due to the relation between these two learning rules, a tempotron-like *ReSuMe* learning rule is proposed. In the proposed remote supervision, the synaptic efficacy between given pre-and postsynaptic neurons does not only depend on the correlated pair.

3.3.3 SpikeProp *Algorithm*

SpikeProp is an error-back propagating learning algorithm [2]. The error to minimize is the mean squared error defined on the spike times of the output neurons and the desired spike times t.

$$E = \frac{1}{2}\Sigma_{j\in O}(t_j^a - t_j^d) \tag{3.5}$$

where t_j^a defines actual spike times of the output neurons, and t_j^d is desired spike times of the output neurons. For each connection k from neuron i to neuron j with weight w_{ij}^k we need to calculate:

$$\Delta w_{ij}^k = -\eta \frac{\partial E}{\partial w_{ij}^k}, \qquad \eta \text{ is learning rate} \tag{3.6}$$

Since t_j is a function of the threshold post-synaptic input x_j which depends on the weight, we expand the derivative and obtain the following equation:

$$\frac{\partial E}{\partial w_{ij}^k} = \frac{\partial E}{\partial t_j}(t_j^a)\frac{\partial t_j}{\partial w_{ij}^k}(t_j^a) = \frac{\partial E}{\partial t_j}(t_j^a)\frac{\partial t_j}{\partial x_j(t)}(t_j^a)\frac{\partial x_j(t)}{\partial w_{ij}^k}(t_j^a) \tag{3.7}$$

For a small enough region around $t = t_j^a$, the function x_j is approximated by a linear function of t, as shown in Fig. 3.5, which is also shown as below:

$$\delta t_j(x_j) = \delta x_j(t_j)/\alpha \tag{3.8}$$

α represents the local derivative of $x_j(t)$ w.r.t. t. Combining the previous results, we then obtain:

$$\Delta w_{ij}^k(t_j^a) = -\eta \frac{y_i^k(t_j^a)(t_j^d - t_j^a)}{\sum_{i\in\Gamma_j}\sum_l w_{ij}^l \partial y_i^l(t_j^a)/\partial t_j^a} \tag{3.9}$$

Fig. 3.5 Relationship between δx_j and δt_j for a small region around $t = t_j^a$

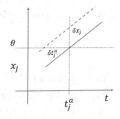

Algorithm 1 *SpikeProp* algorithm

1: Calculate δ_j for all outputs in the final layer
2: For each subsequent layer I, calculate δ_i or all neurons in I
3: For output layer J, adapt w_{ij}^k by $\delta w_{ij}^k = -\eta y_i^k(t_j^a)\delta j$
4: For each subsequent layer I, adapt w_{ij}^k by $\delta w_{hi}^k = -\eta y_h^k(t_i^a)\delta_i$

We define δ_j for the ease of expression, yielding Eq. 3.11, which gives us the weight updating algorithm for neurons in the output layer:

$$\delta_j = \frac{(t_j^d - t_j^a)}{\sum_{i \in \Gamma_j} \sum_l w_{ij}^l \partial y_i^l(t_j^l)/\partial t_j^a} \tag{3.10}$$

$$\Delta w_{ij}^k(t_j^a) = -\eta y_i^k(t_j^a)\delta_j \tag{3.11}$$

For hidden layers, δ_i is defined for $i \in I$ with actual firing times t_j^a:

$$\begin{aligned}
\delta_i &\equiv \frac{\partial t_i^a}{\partial x_i(t_i^a)} \frac{\partial E}{\partial t_i^a} \\
&= \frac{\partial t_i^a}{\partial x_i(t_i^a)} \Sigma_{j \in \Gamma^i} \frac{\partial E}{\partial t_j^a} \frac{\partial t_j^a}{\partial x_j(t_j^a)} \frac{\partial x_j(t_j^a)}{\partial t_i^a} \\
&= \frac{\partial t_i^a}{\partial x_i(t_i^a)} \Sigma_{j \in \Gamma^i} \delta_j \frac{\partial x_j(t_j^a)}{\partial t_i^a} \\
&= \frac{\Sigma_{j \in \Gamma^i} \delta_j \left\{ \Sigma_k w_{ij}^k (\partial y_i^k(t_j^a)/\partial t_i^a) \right\}}{\Sigma_{h \in \Gamma_i} \Sigma_l w_{hi}^l (\partial y_h^l(t_i^a)/\partial t_i^a)}
\end{aligned} \tag{3.12}$$

Thus, we have:

$$\Delta w_{hi}^k = -\eta y_h^k(t_i^a)\delta_i \tag{3.13}$$

The summarized *SpikeProp* algorithm is listed in Algorithm 1:

In Xin and Embrechts' work [50], a momentum term was added to improve convergence and tackle possible occurrence of local minimum:

$$w_{ij}^{k+1} = w_{ij}^k + \eta_k \delta_{ij}^k + \alpha \Delta w_{ij}^{k-1} \tag{3.14}$$

where η_k denotes the learning rate, and α denotes the momentum parameter. Considering that the original version of *SpikeProp* is designed for one spike per neuron, Booij et al. [3] propose a more generic architecture, which contains recurrent connections. This generic architecture can be applied to handle multiple spikes per neuron at a time. In Mckennoch et al.'s work [20], a learning-rate adjustment algorithm, called resilient propagation (*RProp*), is applied in combination with *SpikeProp* to accelerate training process. The *RProp* algorithm, first proposed by Riedmiller and Braun [31], performs the weight-update based on the sign of the gradient. Initially, the value of Δ_{ij} is introduced to associate with each single weight, the update of Δ_{ij} follows the rule:

$$\Delta_{ij}(t) = \begin{cases} \eta^+ \times \Delta_{ij}(t-1), & \text{if } \frac{\partial E}{\partial w_{ij}}(t-1) \times \frac{\partial E}{\partial w_{ij}}(t) > 0 \\ \eta^- \times \Delta_{ij}(t-1), & \text{if } \frac{\partial E}{\partial w_{ij}}(t-1) \times \frac{\partial E}{\partial w_{ij}}(t) < 0 \\ \Delta_{ij}(t-1), & \text{otherwise} \end{cases} \qquad (3.15)$$

where $0 < \eta^- < 1 < \eta^+$, indicating that if the sign of the partial derivative remains unchanged w.r.t. the corresponding weight w_{ij} of the error term, Δ_{ij} should be increased to speed up convergence. In contrast, the update-value is decreased as the sign changes, which means the local minimum has been jumped over. The update of weights is performed thereafter as:

$$w_{ij}(t+1) = \begin{cases} w_{ij}(t) - \Delta_{ij}(t), & \text{if } \frac{\partial E}{\partial w_{ij}}(t) > 0 \\ w_{ij}(t) + \Delta_{ij}(t), & \text{if } \frac{\partial E}{\partial w_{ij}}(t) < 0 \\ w_{ij}(t), & \text{otherwise} \end{cases} \qquad (3.16)$$

However, to ensure that each input neuron initially fires and intermediate neurons should subsequently fire, an appropriate selection of parameters, including learning rates and input weights, should be carefully done. Similar to the work in *RProp* that studied weight initialization, Shrestha and Song [39] proposed a learning rate adaptation method, referred to as *SpikePropAD*, by analyzing the issue of weight convergence. They further offered a robust adaptive learning rate (*SpikePropR*), which satisfies both the weight convergence and robust stability conditions [40]. Another modification is to tailor *QuickProp* method to *SpikeProp*, which is under the assumption that the error function can be depicted as an upward open parabola, and the second derivative of the error regarding one weight is independent from others. Based on these assumptions, Newton's method is used to minimize the one-dimensional error function:

$$\Delta w_{ij}(t+1) = \frac{S(t+1)}{S(t) - S(t+1)} \Delta w_{ij}(t) \qquad (3.17)$$

where $S(t)$ denotes the partial derivative with regards to the corresponding weight w_{ij}. In *QuickProp* algorithm, the current weight change depends on the previous weight change, and the error minimum can be slowly reached except for the large step size during the training phase [20].

In the original *SpikeProp*, each connection contains a fixed number of delayed synaptic terminals (1–16 ms), and only the weights are trained. Such specifications, or constraints, can be relaxed by allowing delays to be trained during learning, thus reducing the number of synaptic terminals and weights [36].

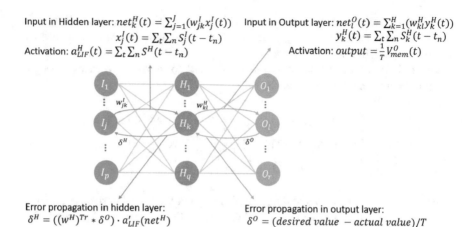

Input in Hidden layer: $net_k^H(t) = \sum_{j=1}^{J}(w_{jk}^I x_j^I(t))$

$x_j^I(t) = \sum_t \sum_n S_j^I(t - t_n)$

Activation: $a_{LIF}^H(t) = \sum_t \sum_n S^H(t - t_n)$

Input in Output layer: $net_l^O(t) = \sum_{k=1}^{H}(w_{kl}^H y_k^H(t))$

$y_k^H(t) = \sum_t \sum_n S_k^H(t - t_n)$

Activation: $output = \frac{1}{T} V_{mem}^O(t)$

Error propagation in hidden layer:

$\delta^H = ((w^H)^{Tr} * \delta^O) \cdot a_{LIF}'(net^H)$

Error propagation in output layer:

$\delta^O = (desired\ value - actual\ value)/T$

Fig. 3.6 Approximate derivative method

3.3.4   Approximate Derivative Method (ADM)

An alternative way to use spike-based backpropagation is to approximate the activation of IF neuron (no leak in the membrane potential) at first step, then a leak factor is introduced to compensate the leaky effect [18]. Figure 3.6 depicts the bidirectional backpropagation of this method.

3.4   Unsupervised Learning

STDP is a biological process that characterizes the synaptic plasticity for adjusting the strength of connections between neurons in the brain [9, 22]. Under the STDP learning mechanism, if an input spike to a postsynaptic neuron precedes an output spike from the neuron, the synapse through which the input was received is strengthened, enabling it to contribute more to the spiking of the postsynaptic neuron in the upcoming period. However if the input spike arrives after the output spike, the synapse through which it was received is weakened, reducing its contribution in the upcoming period. According to the Hebbian rule, synapses increase their efficacy if they persistently participate in the firing of a postsynaptic neuron. Figure 3.7 shows the change of synaptic weight relating to the temporal difference between a pair of presynaptic and postsynaptic spikes. The change in weight of a synapse can be expressed as:

$$\Delta w = \begin{cases} A_+ e^{+\Delta t/\tau_+}, & \Delta t < 0, A_+ > 0 \\ A_- e^{-\Delta t/\tau_-}, & \Delta t > 0, A_- < 0 \end{cases} \qquad (3.18)$$

Fig. 3.7 Basic STDP learning rule

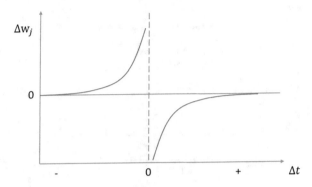

where $\Delta t = t_{pre} - t_{post}$, denoting the time difference between presynaptic and its postsynaptic spike, A_+ and A_- denote the learning rate depending on the synaptic weight. τ_+ and τ_- are the time constants [5].

3.4.1 Pair-Based STDP Learning Rule

The pair-based STDP rule (see Fig. 3.8) considers the time difference of a pair of pre- and post- synaptic spikes, in which a potentiation potential r_1 and a depression potential o_1 are updated iteratively w.r.t. the differential equations below:

$$\frac{dr_1(t)}{dt} = -\frac{r_1(t)}{\tau_+}, \text{ if } t = t_{pre}, \text{ then } r_1 \rightarrow r_1 + 1 \tag{3.19}$$

$$\frac{do_1(t)}{dt} = -\frac{o_1(t)}{\tau_-}, \text{ if } t = t_{post}, \text{ then } o_1 \rightarrow o_1 + 1 \tag{3.20}$$

The potentiating potential r_1 changes whenever a presynaptic spike occurs, and the depression potential o_1 updates whenever there is a postsynaptic spike. The weight change at time when either pre- or post- synaptic spike arrives is then described as:

$$w(t) \rightarrow w(t) - A_2^- o_1(t), \text{ if } t = t_{pre} \tag{3.21}$$

$$w(t) \rightarrow w(t) + A_2^+ r_1(t), \text{ if } t = t_{post} \tag{3.22}$$

where A_2^- and A_2^+ denote the amplitude of weight change controlling the long-term depression (LTD) and the long-term potentiation (LTP) term, respectively.

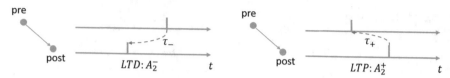

Fig. 3.8 Pair-based STDP learning rule

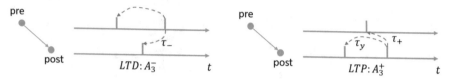

Fig. 3.9 Triplet-based STDP learning rule

3.4.2 Triplet STDP Learning Rule

An observation is that in pair-based rule, the potentiation is exhibited with pre-before-post or post-before-pre timing in the limit of low frequency without taking into account the interaction between the pair of pre-and postsynaptic spikes with others. However, with the increase of frequency, there might rise an additional impact from the presynaptic spikes of the following couple on the postsynaptic spike of the previous pair. This property can not be captured by the standard pair-based model [29]. The pair-based model can be extended to triplet-based STDP, whereby either a pre-post-pre and post-pre-post scheme is formed (Fig. 3.9). The time constant τ_x and τ_y are introduced for the decaying potential between two presynaptic spikes. Accordingly, the update rule of two additional variables r_2 and o_2 can be written as:

$$\frac{dr_2(t)}{dt} = -\frac{r_2(t)}{\tau_x}, \text{ if } t = t_{pre}, \text{ then } r_2 \rightarrow r_2 + 1 \tag{3.23}$$

$$\frac{do_2(t)}{dt} = -\frac{o_2(t)}{\tau_y}, \text{ if } t = t_{post}, \text{ then } o_2 \rightarrow o_2 + 1 \tag{3.24}$$

Therefore, the weight change in triplet STDP learning rule can be extended as follows:

$$w(t) \rightarrow w(t) - o_1(t)[A_2^- + A_3^- r_2(t - \varepsilon)], \text{ if } t = t_{pre} \tag{3.25}$$

$$w(t) \rightarrow w(t) + r_1(t)[A_2^+ + A_3^+ o_2(t - \varepsilon)], \text{ if } t = t_{post} \tag{3.26}$$

where A_3^- and A_3^+ denote the amplitude of the weight change of post successive pre-synaptic pairs and post-synaptic pairs, respectively. Note that the above triplet learning rule is associated with All-to-All interactions. In Nearest-spike interactions,

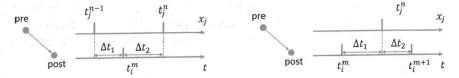

Fig. 3.10 Spike pairing scheme. All-to-All interaction and Nearest-Neighbor interaction scheme

where only the nearest spikes are considered, the update rule for synaptic weights will be modified accordingly (Fig. 3.10). Wang et al. [48] propose the Quadruplet protocol by further taking the interaction between a post-pre pair and a pre-post pair into consideration.

3.4.3 Reward-Modulated STDP Learning

The reward modulated STDP learning (R-STDP) rule extends the unsupervised STDP learning rule by adding sparse external reinforcement signals that can modulate an SNN. The brain to a great extent has been found in physiological experiments to exhibit a reward system using a form of neuromodulator called dopamine (DA) [37]. Dopaminergic neurons in the brain show behaviors akin to rewards by inflecting synaptic plasticity at corticostriatal synapses [16]. This reward mechanism depends on the modulation of dopamine during synaptic adaptation by STDP. The work in [51] shows how the weight changes are determined for R-STDP using the equation.

$$\dot{w} = e \times (d - b) \tag{3.27}$$

where w is the synaptic weight change, e the eligibility trace, d the reward function, and b the baseline. The eligibility traces and reward function can be computed as

$$\dot{e} = -\frac{e}{\tau_e} + \mathrm{STDP}(\Delta t)\delta(t - t_{pre/post}) \tag{3.28}$$

$$\dot{d} = -\frac{d}{\tau_d} + \frac{\delta(t - t_n)}{\tau_d} \tag{3.29}$$

where τ_e is the synaptic eligibility time constant, δ the dirac delta function, and τ_d the neuromodulator concentration time constant.

3.4.4 Other Variants of STDP Learning Rule

An alternative is proposed to model the STDP scheme using only one dynamic variable instead of measuring the time difference in the pair-based model [4]. The idea is to consider only the voltage dependence of the single postsynaptic neuron of membrane $V(t)$ with an integrate-and-fire neuron model. In this model, the updating of the synaptic weight depends on the membrane voltage threshold and a function of postsynaptic spiking activity, which is referred to as the Calcium concentration of a neuron.

3.5 Chapter Summary

The human brain is regarded as a power-efficient learning machine capable of carrying out complex computations while using only little resources. A sophisticated property that makes energy-efficient computation possible is the distinct sparse communication among many spiking neurons. Thus, spiking neural networks gained popularity by incorporating learning. In these neural networks, there are two fundamental modes: Inference and learning. This chapter presented how learning in neuromorphic computing systems is conducted. The learning phase, which minimizes a particular cost function, is a complex process of acquiring the parameters to output the correct inference results. In contrast, inference is computing the output values based on the given input and the network parameters.

References

1. Ben Abdallah A, Dang KN (2021) Toward robust cognitive 3d brain-inspired cross-paradigm system. Frontiers Neurosci 15:795
2. Bohte SM, Kok JN, La Poutré JA (2000) Spikeprop: backpropagation for networks of spiking neurons. In: ESANN, vol 48, pp 17–37
3. Booij O, tat Nguyen H (2005) A gradient descent rule for spiking neurons emitting multiple spikes. Inf Process Lett 95(6):552–558
4. Brader JM, Senn W, Fusi S (2007) Learning real-world stimuli in a neural network with spike-driven synaptic dynamics. Neural Comput 19(11):2881–2912
5. Cai W, Ellinger F, Tetzlaff R (2014) Neuronal synapse as a memristor: Modeling pair-and triplet-based stdp rule. IEEE Trans Biomed Circuits Syst 9(1):87–95
6. Cao Y, Chen Y, Khosla D (2015) Spiking deep convolutional neural networks for energy-efficient object recognition. Int J Comput Vis 113(1):54–66
7. Diehl PU, Neil D, Binas J, Cook M, Liu SC, Pfeiffer M (2015) Fast-classifying, high-accuracy spiking deep networks through weight and threshold balancing. In: 2015 International joint conference on neural networks (IJCNN). IEEE, pp 1–8
8. Florian RV (2008) Tempotron-like learning with resume. In: International conference on artificial neural networks. Springer, pp 368–375
9. Gerstner W, Kempter R, Van Hemmen JL, Wagner H (1996) A neuronal learning rule for sub-millisecond temporal coding. Nature 383(6595):76–78
10. Gütig R, Sompolinsky H (2006) The tempotron: a neuron that learns spike timing–based decisions. Nat Neurosci 9(3):420–428
11. Han B, Srinivasan G, Roy K (2020) Rmp-snn: Residual membrane potential neuron for enabling deeper high-accuracy and low-latency spiking neural network. In: Proceedings of the IEEE/CVF conference on computer vision and pattern recognition, pp 13558–13567

12. Hu Y, Tang H, Wang Y, Pan G (2018) Spiking deep residual network. Preprint. arXiv:1805.01352
13. Hunsberger E, Eliasmith C (2015) Spiking deep networks with LIF neurons. Preprint. arXiv:1510.08829
14. Ikechukwu OM, Dang KN, Abdallah AB (2021) On the design of a fault-tolerant scalable three dimensional NoC-based digital neuromorphic system with on-chip learning. IEEE Access 9:64331–64345
15. Ioffe S, Szegedy C (2015) Batch normalization: Accelerating deep network training by reducing internal covariate shift. Preprint. arXiv:1502.03167
16. Izhikevich EM (2007) Solving the distal reward problem through linkage of STDP and dopamine signaling. Cerebral Cortex 17(10):2443–2452
17. Krizhevsky A, Sutskever I, Hinton GE (2012) Imagenet classification with deep convolutional neural networks. In: Advances in neural information processing systems, pp 1097–1105
18. Lee C, Sarwar SS, Panda P, Srinivasan G, Roy K (2020) Enabling spike-based backpropagation for training deep neural network architectures. Frontiers Neurosci 14, 119
19. Masquelier T, Thorpe SJ (2007) Unsupervised learning of visual features through spike timing dependent plasticity. PLoS Comput Biol 3(2), e31
20. McKennoch S, Liu D, Bushnell LG (2006) Fast modifications of the spikeprop algorithm. In: The 2006 IEEE international joint conference on neural network proceedings. IEEE, pp 3970–3977
21. Mikolov T, Kombrink S, Burget L, J Černockỳ, Khudanpur S (2011) Extensions of recurrent neural network language model. In: 2011 IEEE international conference on acoustics, speech and signal processing (ICASSP). IEEE, pp 5528–5531
22. Morrison A, Diesmann M, Gerstner W (2008) Phenomenological models of synaptic plasticity based on spike timing. Biol Cybern 98(6):459–478
23. Nair V, Hinton GE (2010) Rectified linear units improve restricted Boltzmann machines. In: Proceedings of the 27th international conference on machine learning (ICML-10), pp 807–814
24. Natschläger T, Ruf B (1998) Spatial and temporal pattern analysis via spiking neurons. Network Comput Neural Syst 9(3):319–332
25. Neil D, Pfeiffer M, Liu SC (2016) Learning to be efficient: Algorithms for training low-latency, low-compute deep spiking neural networks. In Proceedings of the 31st annual ACM symposium on applied computing, pp 293–298
26. Nessler B, Pfeiffer M, Maass W (2009) STDP enables spiking neurons to detect hidden causes of their inputs. In: Advances in neural information processing systems, pp 1357–1365
27. Orchard G, Meyer C, R Etienne-Cummings, Posch C, Thakor N, Benosman R (2015) Hfirst: a temporal approach to object recognition. IEEE Trans Pattern Anal Mach Intell 37(10):2028–2040
28. Pérez-Carrasco JA, Zhao B, Serrano C, Acha B, Serrano-Gotarredona T, Chen S, Linares-Barranco B (2013) Mapping from frame-driven to frame-free event-driven vision systems by low-rate rate coding and coincidence processing–application to feedforward convnets. IEEE Trans Pattern Anal Mach Intell 35(11):2706–2719
29. Pfister JP, Gerstner W (2006) Triplets of spikes in a model of spike timing-dependent plasticity. J Neurosci 26(38):9673–9682
30. Ponulak F (2006) Supervised learning in spiking neural networks with resume method. Phd, Poznan University of Technology 46:47
31. Riedmiller M, Braun H (1993) A direct adaptive method for faster backpropagation learning: The RPROP algorithm. In: IEEE international conference on neural networks. IEEE, pp 586–591
32. Roy K, Jaiswal A, Panda P (2019) Towards spike-based machine intelligence with neuromorphic computing. Nature 575(7784):607–617
33. Rueckauer B, Liu SC (2018) Conversion of analog to spiking neural networks using sparse temporal coding. In: 2018 IEEE international symposium on circuits and systems (ISCAS). IEEE, pp 1–5

34. Rueckauer B, Lungu IA, Hu Y, Pfeiffer M, Liu SC (2017) Conversion of continuous-valued deep networks to efficient event-driven networks for image classification. Frontiers Neurosci 11:682
35. Scherer D, Müller A, Behnke S (2010) Evaluation of pooling operations in convolutional architectures for object recognition. In International conference on artificial neural networks. Springer, pp 92–101
36. Schrauwen B, Van Campenhout J (2004) Extending spikeprop. In 2004 IEEE international joint conference on neural networks (IEEE Cat. No. 04CH37541). IEEE, vol 1, pp 471–475
37. Schultz W (1998) Predictive reward signal of dopamine neurons. J Neurophysiol 80(1):1–27
38. Sengupta A, Ye Y, Wang R, Liu C, Roy K (2019) Going deeper in spiking neural networks: VGG and residual architectures. Frontiers Neurosci 13:95
39. Shrestha SB, Song Q (2015) Adaptive learning rate of spikeprop based on weight convergence analysis. Neural Netw 63:185–198
40. Shrestha S, Song Q (2017) Robust learning in spikeprop. Neural Netw 86:54–68
41. Simonyan K, Zisserman A (2014) Very deep convolutional networks for large-scale image recognition. Preprint. arXiv:1409.1556
42. Srivastava N, Hinton G, Krizhevsky A, Sutskever I, Salakhutdinov R (2014) Dropout: a simple way to prevent neural networks from overfitting. J Mach Learn Res 15(1):1929–1958
43. Tang J, Yuan F, Shen X, Wang Z, Rao M, He Y, Sun Y, Li X, Zhang W, Li Y, et al (2019) Bridging biological and artificial neural networks with emerging neuromorphic devices: fundamentals, progress, and challenges. Adv Mater 31(49):1902761
44. Vu TH, Ikechukwu OM, Abdallah AB (2019) Fault-tolerant spike routing algorithm and architecture for three dimensional NoC-based neuromorphic systems. IEEE Access 7:90436–90452
45. Vu TH, Murakami Y, Abdallah AB (2019) Graceful fault-tolerant on-chip spike routing algorithm for mesh-based spiking neural networks. In: 2019 2nd International conference on intelligent autonomous systems (ICoIAS), Singapore, February 2019
46. Vu TH, Murakami Y, Abdallah AB (2019) A low-latency tree-based multicast spike routing for scalable multicore neuromorphic chips. In: ACM 5th international conference of computing for engineering and sciences, Hammamet, Tunisia, July 2019
47. Vu TH, Okuyama Y, Abdallah AB (2019) Comprehensive analytic performance assessment and k-means based multicast routing algorithm and architecture for 3d-NoC of spiking neurons. ACM J Emerg Technol Comput Syst 15(4):1–28
48. Wang HX, Gerkin RC, Nauen DW, Bi GQ (2005) Coactivation and timing-dependent integration of synaptic potentiation and depression. Nat Neurosci 8(2):187–193
49. Wu J, Chua Y, Zhang M, Yang Q, Li G, Li H (2019) Deep spiking neural network with spike count based learning rule. In 2019 International joint conference on neural networks (IJCNN). IEEE, pp 1–6
50. Xin J, Embrechts MJ (2001) Supervised learning with spiking neural networks. In: IJCNN'01. International joint conference on neural networks. Proceedings (Cat. No. 01CH37222). IEEE, vol 3, pp 1772–1777
51. Yan H, Liu X, Huo H, Fang T (2019) Mechanisms of reward-modulated STDP and winner-take-all in bayesian spiking decision-making circuit. In: Neural information processing. Springer International Publishing, pp 162–172
52. Yu AJ, Giese MA, Poggio TA (2002) Biophysiologically plausible implementations of the maximum operation. Neural Comput 14(12):2857–2881
53. Zambrano D, Bohte SM (2016) Fast and efficient asynchronous neural computation with adapting spiking neural networks. Preprint. arXiv:1609.02053
54. Zhang L, Zhou S, Zhi T, Du Z, Chen Y (2019) Tdsnn: From deep neural networks to deep spike neural networks with temporal-coding. In: Proceedings of the AAAI conference on artificial intelligence, vol 33, pp 1319–1326

Chapter 4
Emerging Memory Devices for Neuromorphic Systems

Abstract To design a neuromorphic system in hardware, it is imperative to develop artificial neurons that mimic biological neurons and artificial synapses that emulate biological synapses. Recently, numerous efforts have been made to realize artificial synapses using post-CMOS devices, including resistive random access memory (ReRAM), ferroelectric field-effect transistor (FeFET), phase change memory devices, magnetoresistive random access memory (MRAM), etc. A non-CMOS neuron based on emerging devices has also been investigated. This chapter discusses the major emerging memory technologies that promise neuromorphic computing and highlight some recent significant progress on device studies. The advantages and challenges for each device technology are also discussed.

4.1 Introduction

Neuromorphic computing systems are generally built with thousands or even millions or neurons [2, 12]. As a result, neuromorphic systems' parameters and temporal values are too large to be stored locally. Such a large data must be accessible to support inference, learning, and debugging. However, using memory for storing or accessing such data is one of the key design challenges of current neuromorphic systems. As state-of-the-art systems typically try to mimic the response time of biological systems, a consensus is to operate the neuron in serial mode at higher clock speeds [1, 7, 10]. Hence, storing neuron parameters into memory and reloading when computing is unavoidable.

Since a typical memory has a significant area cost, selecting the technology and organization of the memory in a neuromorphic system is a challenging task. With decades of research and development, there is a wide range of available memory technologies. One feature to consider is the trade-off between the area cost, power consumption, read/write speed, and retention period. Conventional computer memory technologies are generally organized in a pyramid shape, as shown in Fig. 4.1 (left). At the top level, the processing engine (processor) can directly access

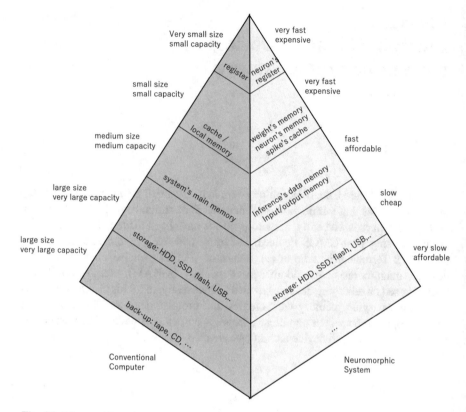

Fig. 4.1 Memory hierarchy

a small static memory (SRAM) known as registers. This type of memory usually has a small capacity and fast access time. A series of D-flip-flops is used as a register that allows nanoseconds access time. However, this type of memory is expensive. As we go from top to down the pyramid, the memory size increases as the area cost become smaller. However, the response time also increases. Across the spectrum of memory technologies, the faster access time the memory technology has, the higher cost per bit it has. To have a greater capacity, we must sacrifice some response time.

Figure 4.2a shows the biological neuron model where dendrites receive inputs from upstream neurons via the synapses. Incoming spikes are integrated into the called *soma* as its membrane potential. If the membrane potential crosses the threshold, the neuron generates an outgoing spike to an axon. The axon sends the spike to the downstream neurons via synapses. Figure 4.2b shows an example of a spiking neuron. The input spikes are multiplied (memory reading) with the corresponding weights to have weighted inputs. The membrane potential is accumulated from the weighted inputs and creates an outgoing spike if it is higher than the threshold.

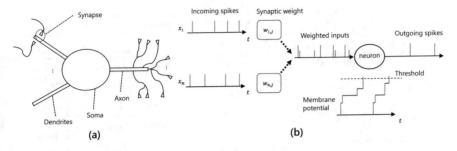

Fig. 4.2 (a) Biological neuron. (b) Spiking neuron

In the spiking neuron models, there are three major parameters than need to be stored (memorized): (1) incoming spikes; (2) synaptic weights, and (3) neuron's internal parameters (membrane potential, threshold, etc.).

Inspired by the hierarchical system of computer memory, Fig. 4.1 (right) depicts the memory hierarchy for neuromorphic systems with a similar shape. The processing engine of neuromorphic systems known as neuron can only access its local information and a certain amount of inputs simultaneously. This information is usually stored in local *neuron's registers* and directly accessed by the processing engine (neuron) The weights of the synapses will be loaded from the lower level of the hierarchy to the local memory. The *weights* are typically stored in a second-tier memory level which requires one or several clock cycles. Moreover, if the physical neuron is shared, the *neuron's parameters* are also stored in a second-level memory. The final part is the *spike's cache* for the incoming and outgoing spikes. The third-tier memory level can be used for loading pre-trained weights, storing input and output spikes or for system configuration.

4.2 Memory Technology

Memory has been one of the essential parts of computing systems. This section covers the major storage technologies used in neuromorphic computing systems.

The basic element of semiconductor memory is the memory cell. Despite multiple variations of memory technologies, all semiconductor memory cells follow the same following properties:

- They exhibit in states or levels, typically two binary values: 0 and 1. There are multiple-level cell technologies that can represent more than two levels.
- They can be written into to set the state and capable of being read to sense the state.
- A typical memory cell has two control inputs: *select* and *control* and two data inputs/outputs (or one in a duplex mode). The *control* input signals the direction

Table 4.1 The taxonomy of memory technologies with key design parameters

Technology	Cell size (F^2)	Write endurance	Speed (R/W)	Leakage power	Dynamic energy (R/W)	Retention period
Register	2200–3500	10^{16}	Extremely fast	Very high	Low	Voltage applied
SRAM	120–200	10^{16}	Very fast	High	Low	Voltage applied
eDRAM	60–100	10^{16}	Fast	Medium	Medium	30–100 μs
STT-RAM	6–50	4×10^{12}	Fast/slow	Low	Low/high	Years
RRAM	4–10	10^{11}	Fast/slow	Low	Low/high	Years
PCM	4–12	$10^8–10^9$	Slow/very slow	Low	Medium/high	Years
DWM	≥2	10^{16}	Fast/slow	Low	Low/high	Years
Flash (NAND)	1–4	10^4	Very slow	Very low	Low	Years

of the data to be read from or written to a memory cell. The *select* input indicates whether the *control* is for the current cell.

Table 4.1 lists the taxonomy of memory technologies. The register is usually used as local memory for processing engines among the memory technologies due to its breakneck speed (nanosecond access time) and compatibility with CMOS technology. The register allows digital neurons to access information (i.e., membrane potential, threshold, input weight) directly to emulate neuron functions. This allows the register to be at the top level of the memory hierarchy. However, the register's high area cost limits the neuromorphic system from having higher capacity. With F as the feature size of the technology, a register can take around 2200–3500 of F^2, which is 15–20× the size of an SRAM cell. This makes the register become the largest in terms of cell size among other technologies. From the second level of memory, other technologies with smaller cell sizes are considered.

Figure 4.3 depicts the operation and organization of a memory system. Most commonly, memory cells are organized in a 2D array, and the accessing address is split into a row address and a column address. If one row is selected, the value of the row can be read or written by providing the control and data signals. This organization is commonly used for embedded memory. Other memory technologies such as HDD, CD, DVD are organized in serial and can be accessed in different manners.

4.2.1   SRAM

Static random-access memory (SRAM) is one of the most popular conventional memory technologies in the semiconductor. A typical embedded SRAM cell is made up of six transistors, as shown in Fig. 4.4. SRAM cells only use the transistors, and

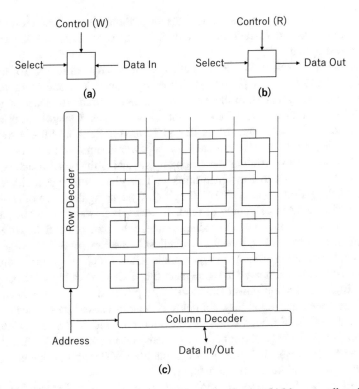

Fig. 4.3 General organization of a memory: (**a**) Memory cell write, (**b**) Memory cell read, (**c**) 2D array of memory cell

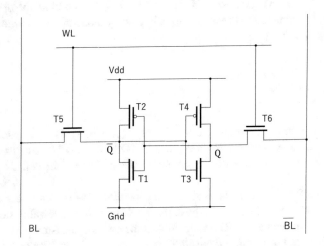

Fig. 4.4 A six transistors (6T) SRAM cell

SRAM arbitration logics use the same gates as in digital design. Therefore, SRAM is fully compatible with CMOS implementation. SRAM cell holds the value as long as power is supplied to it.

Figure 4.4 is a typical 6T SRAM cell. Four transistors (T1, T2, T3, and T4) maintain a stable logic (as Q and $\overline{Q}$). If the logic state Q is one (high voltage), the $\overline{Q}$ is zero (low voltage). In this state, the T1 and T4 transistors are on, and the T2 and T3 transistors are off. If the logic state Q is zero (low voltage), the transistors T1 and T4 are off, and the transistors T2 and T3 are on. For both SRAM states, the transistors maintain their logic level as long as power is supplied to them.

SRAM also supports reading and writing by switching the two transistors T5 and T6. To read, WL is set to one which turns on T5 and T6, which pull either BL or $\overline{BL}$ to zero depending on the value of Q and $\overline{Q}$. By sensing the value of BL or $\overline{BL}$, the value of the SRAM cell could be read. To write, WL is also set to one, and BL and $\overline{BL}$ will be driven with proper value to change the state of the SRAM cell. This forces the four-state holding transistors to adjust to the new state.

The implementation of SRAM is compatible with CMOS design and is supported in most CAD tools. Compared to the register file, SRAM has a higher density (only six transistors) and lower power (no clock, only power supply). The advantages of SRAM are speed and reliability. Its delay is mainly caused by connected wires, and its mature CMOS design could be reliable for around 10^{16} accesses. Its major drawback is high leakage current and low density when compared to other memory technologies. However, it is the most compatible CMOS design, which could be easily integrated without any modification.

In [16], an 8T SRAM design allows both row and column access for learning efficiently in the neuromorphic system. Typical SRAM cells can only be written and read in rows. By adding two more transistors for the transpose reading and writing, the 8T SRAM can allow column access. This can emulate the operation of the crossbar for connecting the presynaptic to post-synaptic neuron as each cell acts like a binary weight synapse.

4.2.2   eDRAM

Embedded dynamic random-access memory (eDRAM) is a capacitor-based dynamic RAM that can be integrated into CMOS. The eDRAM cell could be a conventional 1T1C (1-transistor 1-capacitor) design or a gain cell as shown in Fig. 4.5.

A 1T1C eDRAM cell stores its state in a capacitor C as shown in Fig. 4.5a. The presence and absence of charge in the capacitor C represents the value stored in the eDRAM cell. To read an eDRAM, the WL turns on the transistor, and the voltage of capacitor C is taken into a device called a sense amplifier. The sense amplifier detects the value of eDRAM cell by comparing the voltage with the threshold voltage. To write the eDRAM cell, WL turns on the transistor and the wire BL is connected to either ground or high voltage to change the value of the cell. Because

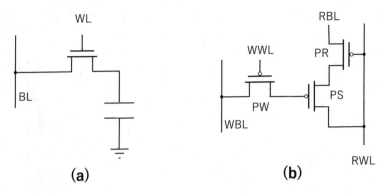

Fig. 4.5 eDRAM cell design: (**a**) 1T1C, (**b**) gain cell

the capacitor charge is leaky by its nature, eDRAM requires periodic refreshing to maintain the state of the eDRAM cell. A typical retention method of eDRAM is to read a row, capture the row's values, and write the value back to the row. The gain cell design in Fig. 4.5b does not use any capacitor. Instead, the transistor will hold the value of the cell. This allows eDRAM to compatible with CMOS design as no capacitor is used. However, this significantly increases the size as three transistors are used.

eDRAM is likely to have less area cost than SRAM, thanks to its simple cell design. However, eDRAM requires a retention period to preserve its value because the storage charge gradually leaks during operation. Also, 1T1C eDRAM is not compatible with CMOS design (an extra process is needed), and reading may cause a loss of high value. Refreshing signal design is complicated for eDRAM due to the high fan-out and skewing phenomenon (similar to clock design).

4.2.3 STT-RAM

Spin-transfer torque RAM (STT-RAM) [4] is a magnetic RAM in which a cell consists of a magnetic tunneling junction (MTJ). MJT consists of two ferromagnets (one is free, one is fixed) separated by a thin insulator. The insulator is thin enough to allow quantum tunneling (electrons jump through the insulator). Depending on the relative magnetization directions of these two ferromagnet layers, the MTJ is either low-resistive (parallel) or high-resistive (anti-parallel). Therefore, STT-RAM works like a non-volatile device. At a high-resistive state, the voltage of BL is 0V due to the high resistance of the MTJ. MTJ allows current go through it to drive the BL close to the supply voltage at a low-resistive state.

Because of its simple design, STT-RAM has a small area when compared to SRAM or eDRAM. The structure of a STT-RAM cell is shown in Fig. 4.6. However, the relative magnetization direction states take a long time to change, making the

Fig. 4.6 A STT-RAM cell

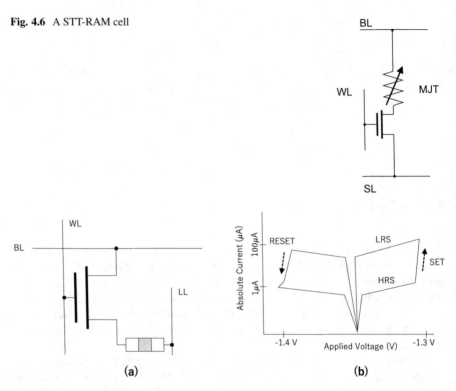

(a)   (b)

Fig. 4.7 RRAM cell: (**a**) Schematic. (**b**) I–V characteristics curve of a HfO_x RRAM cell [17]. Current is in absolute value. Readers may be more familiar with the I–V characteristics of memristor

writing process slow and consumes more power. In the rest of this chapter, STT-SRAM is classified as non-volatile memories (NVMs). Beside SRAMs, resistive NVMs are very common in the neuromorphic system as they are small, and can perform in-memory computing.

4.2.4 RRAM and Resistive Crossbar

Resistive random-access memory (RRAM or ReRAM) can generally denote all memory technologies that rely on the resistance change to store the information [18]. In this chapter, we use the term RRAM for memristor-based memory. Other mentioned memory technologies such STT-RAM and PCM also change the cell resistance for changing the state.

Figure 4.7a shows the 1T1R RRAM cell. Writing is based on the voltage of LL and WL. Reading is based on the voltage of BL. As we can see, this architecture is similar to eDRAM, which offers low area costs. The major advantages of RRAM

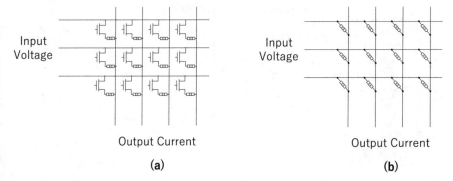

Fig. 4.8 Resistive crossbar design: (**a**) 1T1R. (**b**) 1 0T1R

are its low area cost, low power, and easy integration of CMOS technologies. For neuromorphic applications, a crossbar using memristors could be used with a similar principle.

The principle of RRAM is to apply a certain voltage to the cell, which drives it to set (low resistance) or reset mode (high resistance). Figure 4.7b show the I–V characteristics of an RRAM cell. At the beginning state, the device is at HRS (high-resistive state), making the current less than $1\,\mu A$. A HfO_x RRAM, by applying a bias voltage 1.3 V between two terminals (P and N), generates more oxygen vacancies, spawning more conductive filaments. Then, the RRAM cell change to LRS (low-resistive state). To reset the cell to HRS, the applied voltage is less than -1.3 V.

Figure 4.8 shows the resistive crossbar crossbar, which shares the same principle with RRAM. There are two options: (1) use the transistor as in Fig. 4.7a or (2) the transistor is removed to have 0T1R cells [18]. For the reading process, reading voltage is applied as similar to WL. The BL is read. Depend on the resistance of the memristor; the output voltage could be 0 or near WL voltage. Note that there is a leakage current due to the lack of the transistor. The writing process is done by applying a voltage between two terminals of the memristor. Due to variation and hard to control leakage current under high-density RRAM, the design of 1T1R is more preferable.

4.2.5 Phase Change Memory

One of the most current advanced memory technology is Phase Change Memory (PCM) [5]. PCM is based on the property of certain materials, such as $Ge_2Sb_2Te_5$, which exhibit differences in resistivity in their two phases: crystallized and amorphous. In a PCM device, a small amount of one of the material is put between two metal terminals as shown in Fig. 4.9a. To program the PCM device, a pulse of SET

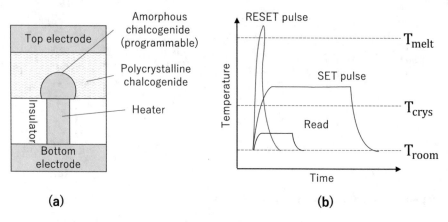

Fig. 4.9 Phase change memory: (**a**) A cross-section image of a mushroom-type PCM device. (**b**) The programming pulses and the resulting relative temperature for RESET, SET, and read operation in PCM

or RESET as in Fig. 4.9b have to be used. While the RESET pulses increase the size of the amorphous region, the SET pulses reduce the size of the area to have more crystallized material. As a result, the conductance of the device will be changed.

PCM is mature and has been investigated for neuromorphic systems [13, 14]. It has some strong points on cell size (4–$12\,F^2$), write endurance, and power consumption. However, it has a latency problem. A typical SET phase is done using partial pulses in a period of $100\,ns$ (50% duty cycle) of $90\,\mu A$ [14] requires hundred of seconds to change phase. Moreover, it also has some deviations in conductance evolution as it can drift the accuracy of the neuromorphic system.

4.2.6 Other Memory Technologies

Table 4.1 shows other memory technologies, such as DWM (domain wall memory), Flash, HDD. While Flash is slow and unsuitable for embedded memories, DWM is immature and should be investigated comprehensively. Spintronics is another advanced technology for neuromorphic systems [11].

Among the memory technologies, SRAM is the most readily available with low power, reasonable area cost, and mature design. eDRAM has a lower cost. However, it consumes high power due to its retention. STT-RAM, PCM, and RRAM is a promising solution with low area cost and low power consumption. For FPGAs or conventional ASIC chips, SRAM is the proper solution, while STT-RAM, PCM, and RRAM are suitable for prototype solutions.

4.3 Memory Organization

The last section has discussed memory cell technologies. This section covers the organization of memory in general semiconductor systems.

Figure 4.10 shows a typical organization of semiconductor memory. Note that with eDRAM, refresh counter and refresh circuitry is necessary to maintain the logic level in the cells. Logically, a semiconductor memory consists of a 2D array of $M \times N$ cells (M rows and N columns). Various arrangements are possible for the same number of bits (cells) in the memory. If the number of the columns is the accessing bit-width (word's width), no column decoder is needed, and the output of the memory access goes straight to the data output buffer. If the accessing bit-width is smaller than N, a column decoder is needed to extract the data out of N bits.

The address for memory is typically used for accessing a word (i.e., a word is typically 32 bits). For instance, a 1-Mbit memory of 32-bit words can be organized by an array of 1024×1024. To access the memory, a 10-bits row address is needed. After a row is read from the memory array, a 5-bits column address is used to determine the reading word. The output is cached in a data output buffer and later sent to the reading block. In the writing process, data is accessible in the same bit line wires to change the value of the row. If only one word is changed in the whole row, the row is read, and the word is replaced. The new row is later rewritten into the memory. In eDRAM, data can also be sent to the data input buffer to refresh the memory.

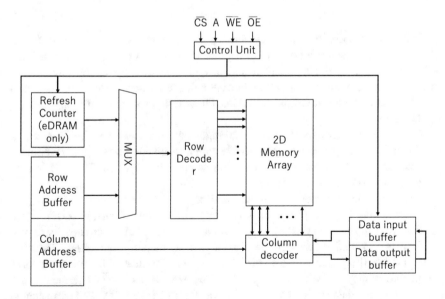

Fig. 4.10 Organization of a semiconductor memory

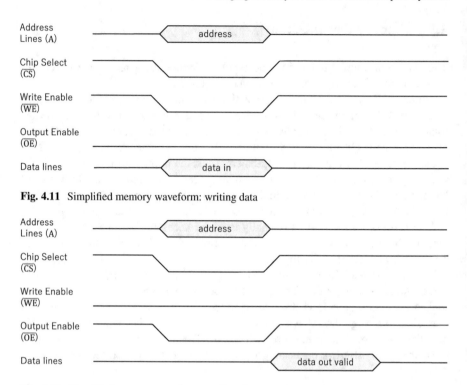

Fig. 4.11 Simplified memory waveform: writing data

Fig. 4.12 Simplified memory waveform: reading data

The reading circuitry can vary among memory technologies. For instance, eDRAM requires sense amplifiers to capture the correct value of the cell. SRAM value is already in the logic level and can be sent directly to the data output buffer. Resistive memory will output current and need a conversion from current to voltage to detect the state of the cell. Also, the writing process is different between technologies.

As we mentioned earlier, various arrangements are possible. For instance, a $32,768 \times 32$ memory array is also possible with a 15-bit row address and no column address. Dividing into several arrays such as four arrays of $32,768 \times 8$ is also possible to provide similar functions.

The I/O ports of a memory block could be: (1) row and column address (A), (2) select signal ($\overline{CS}$), (3) write enable ($\overline{WE}$), and (4) output-enable ($\overline{OE}$). Other fundamental I/O such as ground, supply voltage, reset, and clock are not displayed.

Figures 4.11 and 4.12 show two examples of memory operation. The first operation is writing data with the chip select ($\overline{CS}$) and write enable ($\overline{WE}$) signals re set to zero as they are active at a low level. The output-enable ($\overline{OE}$) signal is high. The address and data arrive at the same time as the chip select ($\overline{CS}$) and write enable ($\overline{WE}$) signals. The memory cell with the address will be written with the content in data lines.

The reading operation is different, with the chip select ($\overline{CS}$) and output enable ($\overline{OE}$) signals active at zero. The address arrives at the same time as the above two signals. The valid data out will appear late after the output enable is received. The typical delay is one clock cycle in most designs.

4.4 Memory for Neuromorphic Systems

Neuromorphic systems typically need to store three major types of data: spikes, neuron states, and weights. While spikes are time-step-based information that needs to be for synchronization, other data types are organized depending on the neuron structure. Spikes could be stored in either SRAM or registers as they can be read directly and converted to synapse array address. Spike memory can be designed in either first-in first-out (FIFO) queue structure or with sorting/scheduling mechanism.

Since the neuron can be either a digital or an analog model, storing and loading the state must be adapted. On the other hand, operating serial or parallel neurons also leads to different storing and loading strategies. The final part of the memory organization is the weight memory, which is the most important because the number of synapses is enormous in any neural network system. In the following parts of this chapter, different neuron designs and operations will be covered with the equivalent memory organization.

4.4.1 Neuron State Memory

As we mentioned earlier, there are two type of neurons: (1) *analog* and (2) *digital* (Fig. 4.13). In principle, both analog and digital neuron has two major types of information to store: *membrane potential* and *threshold*. The in situ learning system might store more information for the learning purpose. The digital neurons' information is already in the binary format and stored in registers for computing purposes. Therefore, loading and storing them is similar to the normal caching mechanism. On the other hand, the analog values of the analog neuron's threshold and memory potential are difficult to store. These values must be digitalized with an ADC and restored later with a DAC.

Furthermore, neurons can be operated in: (1) *serial*: share the physical neuron for multiple neurons' calculations, and (2) *parallel*: each neuron has its physical computing unit. In the parallel model, the state of the neuron could stay in registers of the physical hlneuron, and never need to be reloaded during computation [6]. There is no memory needed unless for loading at the initial state. With the serial one, calculating in time multiplexing is utilized [1, 8, 9]. The parameters are stored in local memory and loaded serially to update the value. The model is illustrated in

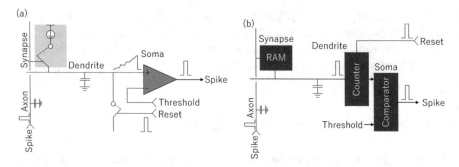

Fig. 4.13 Analog and digital silicon neurons. (**a**) Analog implementation: incoming spikes on the vertical wire (axon) meter charge (synapse) onto the horizontal wire (dendrite), whose capacitance integrates the charge. The comparator (soma) compares the resulting voltage with a threshold and triggers an outgoing spike when the threshold is exceeded. The capacitor is then discharged (reset) and the cycle starts over. (**b**) Digital implementation: a counter is incremented (dendrite) each time a 1 is read out of a bit cell (synapse), triggered by the incoming spike (axon). The counter's output is compared (soma) with a digitally stored threshold and a spike is triggered when it is supra threshold. The counter is then reset and the cycle starts over

Fig. 4.14. The neuron ID is started with 0 and will be looped through all N serial neurons. In each step, after completing the computation of a neuron, the parameters are written back to parameter memory. The basic principle for this time multiplexing is described in Fig. 4.14b:

1. Loop neuron ID $i = 0$ to N−1 (N: number of neurons).
2. Load neuron ID i state (membrane potential, threshold, other parameters).
3. Calculate the neuron operation (integrate, leak, fire).
4. Save neuron ID i to memory.
5. Increase neuron ID i.
6. Loop until the last neuron complete.

The structure of the neuron is shown in Fig. 4.14c where it consists of N rows. Each row consists of the parameters for a neuron in the cluster.

The primary reason for serializing neuron operation is the limitation on weight access. Most of the works are based on SRAM design which is preferred to have a single port R/W for less complexity. Therefore, it is challenging to access multiple weights at the same time.

4.4.2 Synapse Memory

In the memory of neuromorphic systems, the synapses memory is the central part as they occupy most of the storing memory. State of the art neuromorphic memory system can be classified into SRAM-based synapses and non-volatile memory synapses. The SRAM method uses the Static Random Access Memory [4] tech-

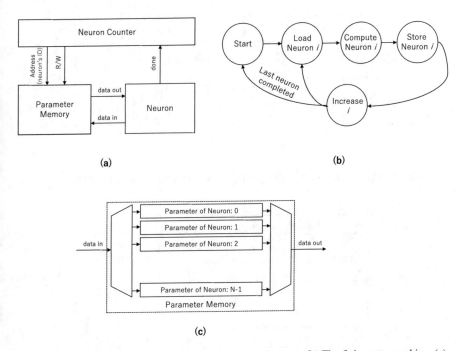

Fig. 4.14 The *serial* neuron model. (**a**) The model architecture. (**b**) The finite state machine. (**c**) The parameter structure

nology, which is well mature, fast, and ready to fabricate with CMOS technology. On the other hand, non-volatile memories (NVMs) are rising as the new key technology to implement neuromorphics system. The general approach for non-volatile is to use a resistive crossbar with multiple conductance levels to mimic the matrix-vector multiplication of neural networks. While SRAM is faster and ready to be fabricated, NVMs are low power and much smaller, allowing large-scale neuromorphic systems. This section covers both types of memory technology and how a neuromorphic system can be designed around these technologies.

4.4.2.1 SRAM Synapse Memory

Similar to neuron state memories, there are *serial* and *parallel* models of synapse memory. In the *parallel* model, each neuron must have its synapses SRAM and perform the weight reading simultaneously. On the other hand, as the *serial* model shares the physical neuron for several emulated ones, having multiple weight memory is inefficient as they are not being used at the same time. The overall model is illustrated in Fig. 4.15 where each neuron is connected directly to its weight memory. Each row of the weight memory consists of the value of a synapse. Once

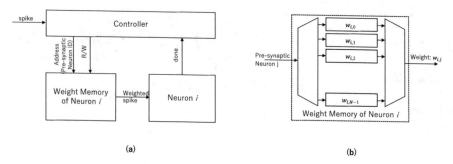

(a)   **(b)**

Fig. 4.15 The *parallel* neuron weight model. (**a**) The model architecture. (**b**) The weight structure

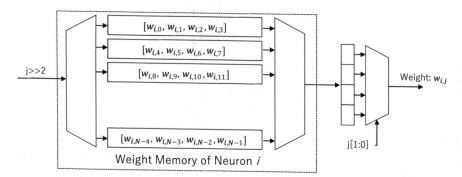

Fig. 4.16 The *parallel* neuron weight memory with merged four weights in a memory row

a spike arrives, the presynaptic neuron ID is extracted (i.e., neuron j) and converted to the memory address.

The SRAM weight memory receives the address corresponding to the ID j of the presynaptic neuron. In the following clock cycle, the weight of the synapse between neuron i and j ($w_{i,j}$) is sent to neuron i for integration

Since the synapse resolution can be low, designers can merge several weights to put them into a memory row. The cell area stays unchanged as the number of bits representing the synapses does not change. However, the complexity for SRAM is reduced due to fewer rows in the SRAM. The structure of this memory model is illustrated in Fig. 4.16. Here, four consecutive weights are merged to establish a memory cell. As a result, the address for a weight between neuron j and neuron i is no longer an address representing the neuron j. Instead, it is the right-shifted value of the address. The remaining two-bit is later used in an additional multiplexer to extract the weight out of four read weights.

The examples of normal and merged weight for parallel neurons are illustrated in Fig. 4.17. With normal weight, the pre-synaptic neuron ID is fed directly to the SRAM to obtain the row of the memory. Figure 4.17a illustrates an array of pre-synaptic spikes [7, 8, 10, 14]. Note that they are not in ascending or descending order which can be problematic for the merged weight model. For example, if the

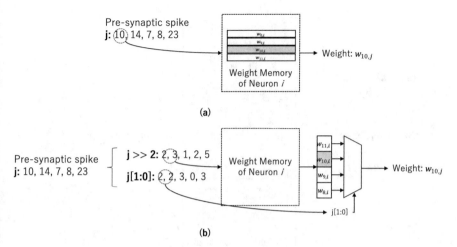

Fig. 4.17 The *parallel* neuron weight memory operation: (**a**) separated weight, (**b**) merged weight

address of the first pre-synaptic neuron $j = 10$, is fed to the weight memory, the following cycle of SRAM output will bring the weight between neuron $j = 10$ and neuron i. With the same array of pre-synaptic neuron ID, the merged weight model split the address into two field: (1) the SRAM address ($j >> 2$) that helps extract a group of 4 weights; and (2) the selection signal ($j[1:0]$) that helps select the correct weight. The pre-synaptic ID $j = 10$ is split into $j >> 2 = 2$ and $j[1:0] = 2$. With the address $j >> 2 = 2$, the weight SRAM outputs a set of four weights: $[w_{8,i}, w_{9,i}, w_{10,i}, w_{11,i}]$. The correct weight $w_{10,i}$ is later extracted by using a multiplexer with a selection signal $j[1:0] = 2$.

Note that with the merged weight model, the system throws away three remaining weights ($w_{8,i}$, $w_{9,i}$, and $w_{11,i}$) if they are not used. Also, if the spike array is unsorted, there is a chance that the system may repeatedly read the same memory row for two consecutive weight. For instance, the spike $j = 10$ and $j = 8$ share the same memory address $j >> 2 = 2$. Without sorting and merging the reading process, the neuromorphic system will read the address $j >> 2 = 2$ twice. The optimal solution is to sort the incoming spikes in ascending or descending order and later merge the shared memory row.

In the *serial* neuron model (Fig. 4.18), having several dedicated memories is unnecessary as this model cannot perform the neuron in parallel. The weight structure here also hlfollows the same manner as the parallel with a tweak. For separated weights, all weights are stored in a single SRAM. Instead of having the presynaptic neuron ID j, the address is $i \times N + j$ (N: number of presynaptic neurons). For N presynaptic and M post-synaptic neurons, the number of banks will be $N \times M$. Reducing the number of banks that can be obtained by merging the neuron. For instance, with small synapses bitwidth, the whole weight of a neuron

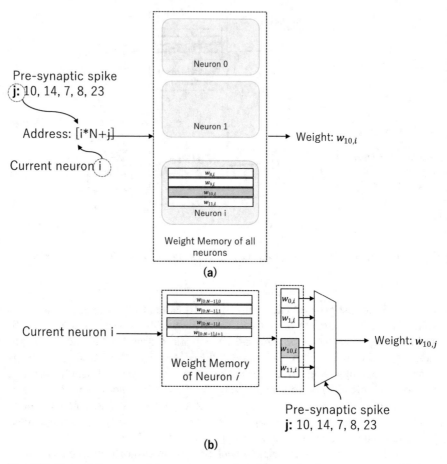

Fig. 4.18 The *serial* neuron weight memory operation: (**a**) normal weight, (**b**) merged weight

can be combined for reading at once [1]. All weights of neuron *i* will be read at once, and a multiplexer will be used to separate the weights later.

In the merged weight mechanism, the state and weight neuron can be merged. In TrueNorth [1], the SRAM word is 410 bits: 256-bits weight (1-bit per synapse), 124-bit neuron state (membrane potential and other parameters), and 30 bit for spike representation. Once a serial neuron is loaded, its parameters and weights are taken from the memory and ready to be computed.

4.4.2.2 Non-volatile Synapse Memory

The first use of NVM is to store the data as same as standard SRAM design. The resistance (or conductance) of the NVM cell can dictate the output current from the crossbar. Consequently, it can be used as a DRAM design (1T1C) where the output voltage is compared to CMOS threshold voltage to obtain the bit value. In this design, the NVM can be used as standard SRAM as each cell stores one bit. Since the cell's conductance can be varied in a wide range, the multiple-level cell can be applied. For instance, having an ADC (or multiple voltage comparators) to measure the output of NVM can convert the conductance to multiple bits value. For RRAM devices, it is reported that a resistive cell can store up to 64-bits [15]. However, this technology is still immature due to its poor retention time at high temperatures, limited endurance, and post-algorithm instability.

For storing data in NVM memory, outputs of NVMs are similar to outputs of SRAM as we previously introduce in Sect. 4.4.2.1. The organization processes can be either serial or parallel. Also, independent and merged weight rows can be used.

4.4.2.3 NVM In-memory Computing

The resistive memory design for synapses can emulate connection and matrix multiplication. Figure 4.19a illustrates the crossbar design using a resistive memory (i.e., 1T1M with M is memristor). By applying input voltages in all lines, the output current is the summary of voltages multiplying the conductance of the resistive devices. In the multiple layer design, the conductance (G_{ij}) is considered as the weight (W_{ij}) between two layers. To have a new connected layer, an activation unit (i.e., RELU in ANN and threshold comparator in SNN) is required to transform the output current to the input voltage of the consequent layer. The crossbar of NVMs can be translated into the synapses and acts for the integration process. The activation unit acts as the threshold in the neuromorphic system, and together it emulates the operation of LIF neuron in Fig. 4.13. The output current for neuron j (I_j) is calculated as the summary of the current provided by all presynaptic neuron voltage (I_{ij}) (the Kirchhoff's law):

$$I_j = \sum I_{ij} \tag{4.1}$$

where I_{ij} is dependent on the applied voltage and the conductance of the NVM cell (as the Ohm's Law):

$$I_{ij} = V_i \times G_{ij} \tag{4.2}$$

Hence, the resistive crossbar can act as matrix multiplication.

$$I_j = \sum V_i \times G_{ij} \tag{4.3}$$

The range of non-volatile memory can be classified either with learning or inference only design. The inference only design can be simple with 1T1R (1 transistor, one resistor) and can translate the matrix multiplication. With the in situ

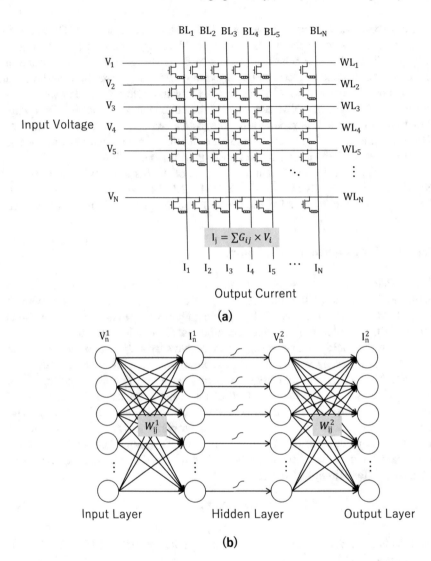

$$I_j = \sum G_{ij} \times V_i$$

(a)

(b)

Fig. 4.19 Schematic for multiple layer neural network using NVM: (**a**) Crossbar for two connected layers. (**b**) Three layer design

learning, in particular the STDP learning, there are long-term depression (LTD) and long-term potentiation (LTP). To mimic the behavior of the relationship between the timing of presynaptic and post-synaptic learning, two resistors can be used for two different types (LTD or LTP). A voltage comparator between the voltage of two resistors is used to determine the output of the synapse. The design is so-called "2-NVM Synapse" [3].

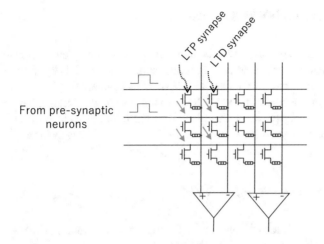

From pre-synaptic
neurons

To post-synaptic neurons

Fig. 4.20 The 2-NVM synapse design

The "2-NVM Synapse" [3] is illustrated in Fig. 4.20. Instead of having a single NVM cell that has the conductance act like the weight of synapses between two neurons, the 2-NVM synapse uses two: one for LTD and one for LTP. The actual weight is considered as the difference between two resistances:

$$W_{ij} = \beta(G_{ij}^{LTP} - G_{ij}^{LTD}) \tag{4.4}$$

where β is the scaling factor, G_{ij}^{LTP} and G_{ij}^{LTD} are the conductance of two NVM cell.

As the output of the resistive crossbar is analog, there are two options to perform the activation as shown in Fig. 4.13. The analog neuron can capture the output voltage as the wire act like a capacitor (virtual ground). The output voltage here can be compared with a threshold voltage to check the firing condition. Once the output voltage is higher than the threshold, an output spike also resets the membrane potential.

In the digital version, the activation unit design depends on the method of applying the input voltages. If the input voltages are applied simultaneously, an ADC can be used to convert the voltage to a digital value and compare it with a threshold value. If the input voltages are applied in serial (one by one), a counter can be used to capturing the incoming weighted spikes. In either case, converting and comparing in the digital domain is undoubtedly need.

4.5 Dynamic NVM Synapse

While NVMs have high density and can provide low power consumption for reading, writing, and holding, the conductance value of NVMs are not always consistent over time. The nature of NVMs is to have drifting in their conductance and can even be stuck at LRS, HRS, ground, or supply voltage. These behaviors make NVM synapses keep changing from the initial state to wear-out time. This section discusses the critical factor in the dynamics of NVM synapses.

4.5.1 Learning Related NVM

Learning in neuromorphic system can be divided in to online (in situ) and offline (ex situ). The ex situ is executed in a software system first, then the calculated weights are loaded to the synapse array. The weights are not adjusted during the loading process. This does not affect SRAM synapses; however, it can be critical in the NVMs system. Also, it is important to maintain the conductance of NVMs during operation. Moreover, even SRAM has potential transient and permanent errors that can change the bit-values. Therefore, the reliability is stringent in *ex-situ* learning.

In in situ learning, the initial weight can be randomized for a new training or pre-trained ones for tuning. The goal of in situ is to minimize the accuracy losses due to deviation of conductance in NVMs. After one training iteration, a pre-inferring process is needed to calculate the difference between expected results and actual results. Then, the tuning processing is done to reduce the error. After tuning, the system needs to verify the conductance of the NVMs to ensure the correctness of the writing process. To reduce the impact of conductance drift, considering only the sign of the weight (negative and positive) in learning is also a possible approach [19]. The conductance drift is critical for both in situ and ex situ learning.

To tune the conductance of the NVM, an external voltage is applied (WL and BL in Fig. 4.13). For a single 1-NVM design, the process of increasing the conductance is called "SET" (or potentiation in neuromorphic). The method of increasing the conductance is called "RESET" (or depression in neuromorphic). In the 2-NVM design, the relative between two NVM cells decides the potentiation or depression of the weight.

4.5.2 Conductance Drift in NVM

As the conductance drift is a typical behavior of the NVM cells, adapting the weights in both in situ and ex-situ learning should be considered. First, the retention of the NVM cell is the first issue to be considered. Unlike SRAM, which needs to be re-fetched after resetting, the NVM cell can maintain the conductance for years.

The second issue is the NVM cell's endurance during the training process (using RESET and SET action). As being summarized in Table 4.1, the RRAM and PCM have the write endurance around 10^{11} and $10^8 - 10^9$, respectively. The failures can be stuck at SET, stuck at RESET, open or stuck-at-ground defects. Third, as the read and write perform analogously, the reading and writing processes can be noisy, leading to inaccurate results. SRAM can deal with this problem by using an Error Correction Code. Here, level-based NVM can certainly be used in companion with ECC; however, fully analog systems are not compatible with such a method.

4.6 Chapter Summary

Recently, numerous efforts have been made to realize artificial synapses using post-CMOS devices, including resistive random access memory (ReRAM), ferroelectric field-effect transistor (FeFET), phase change memory devices, magnetoresistive random access memory (MRAM), and so on. A non-CMOS neuron based on emerging devices has also been investigated. This chapter discussed the major emerging memory technologies that promise neuromorphic computing and highlight some recent significant progress on device studies. The advantages and challenges for each device technology were also discussed.

References

1. Akopyan F, Sawada J, Cassidy A, Alvarez-Icaza R, Arthur J, Merolla P, Imam N, Nakamura Y, Datta P, Nam G, Taba B, Beakes M, Brezzo B, Kuang JB, Manohar R, Risk WP, Jackson B, Modha DS (2015) Truenorth: design and tool flow of a 65 mw 1 million neuron programmable neurosynaptic chip. IEEE Trans Comput-Aid Des Integr Circuits Syst 34(10):1537–1557
2. Ben Abdallah A, Dang KN (2021) Toward robust cognitive 3d brain-inspired cross-paradigm system. Front Neurosci 15:795
3. Bichler O, Suri M, Querlioz D, Vuillaume D, DeSalvo B, Gamrat C (2012) Visual pattern extraction using energy-efficient "2-PCM synapse" neuromorphic architecture. IEEE Trans Electron Dev 59(8):2206–2214
4. Chang M, Rosenfeld P, Lu S, Jacob B (2013) Technology comparison for large last-level caches (L^3CS): low-leakage SRAM, low write-energy STT-RAM, and refresh-optimized eDRAM. In: 2013 IEEE 19th international symposium on high performance computer architecture (HPCA), pp 143–154
5. Close G, Frey U, Breitwisch M, Lung H, Lam C, Hagleitner C, Eleftheriou E (2010) Device, circuit and system-level analysis of noise in multi-bit phase-change memory. In: 2010 international electron devices meeting. IEEE, Piscataway, pp 29–5
6. Dang KN, Abdallah AB (2019) An efficient software-hardware design framework for spiking neural network systems. In: The international conference on internet of things, embedded systems and communications (IINTEC 2019)
7. Davies M, Srinivasa N, Lin T, Chinya G, Cao Y, Choday SH, Dimou G, Joshi P, Imam N, Jain S, Liao Y, Lin C, Lines A, Liu R, Mathaikutty D, McCoy S, Paul A, Tse J, Venkataramanan G, Weng Y, Wild A, Yang Y, Wang H (2018) Loihi: a neuromorphic manycore processor with on-chip learning. IEEE Micro 38(1):82–99

8. Frenkel C, Lefebvre M, Legat JD, Bol D (2018) A 0.086-mm^2 12.7-pj/sop 64k-synapse 256-neuron online-learning digital spiking neuromorphic processor in 28-nm CMOS. IEEE Trans Biomed Circuits Syst 13(1):145–158

9. Frenkel C, Legat J, Bol D (2019) Morphic: a 65-nm 738k-synapse/mm^2 quad-core binary-weight digital neuromorphic processor with stochastic spike-driven online learning. IEEE Trans Biomed Circuits Syst 13:999–1010

10. Furber SB, Lester DR, Plana LA, Garside JD, Painkras E, Temple S, Brown AD (2013) Overview of the spinnaker system architecture. IEEE Trans Comput 62(12):2454–2467

11. Grollier J, Querlioz D, Camsari K, Everschor-Sitte K, Fukami S, Stiles MD (2020) Neuromorphic spintronics. Nat Electron 3(7):360–370

12. Ikechukwu OM, Dang KN, Abdallah AB (2021) On the design of a fault-tolerant scalable three dimensional NoC-based digital neuromorphic system with on-chip learning. IEEE Access 9:64331–64345

13. Joshi V, Le Gallo M, Haefeli S, Boybat I, Nandakumar SR, Piveteau C, Dazzi M, Rajendran B, Sebastian A, Eleftheriou E (2020) Accurate deep neural network inference using computational phase-change memory. Nat Commun 11(1):1–13

14. Nandakumar S, Le Gallo M, Boybat I, Rajendran B, Sebastian A, Eleftheriou E (2018) A phase-change memory model for neuromorphic computing. J Appl Phys 124(15):152135

15. Pérez E, Cristian Zambelli MKM, Olivo P, Wenger C (2019) Toward reliable multi-level operation in rram arrays: improving post-algorithm stability and assessing endurance/data retention. IEEE J Electron Dev Soc 7:740–747

16. Seo J, Brezzo B, Liu Y, Parker BD, Esser SK, Montoye RK, Rajendran B, Tierno JA, Chang L, Modha DS, Friedman DJ (2011) A 45nm CMOS neuromorphic chip with a scalable architecture for learning in networks of spiking neurons. In: 2011 IEEE custom integrated circuits conference (CICC), pp. 1–4

17. Tosson AMS, Yu S, Anis MH, Wei L (2018) Proposing a solution for single-event upset in 1T1R RRAM memory arrays. IEEE Trans Nucl Sci **65**(6), 1239–1247

18. Yang JJ, Strukov DB, Stewart DR (2013) Memristive devices for computing. Nat Nanotechnol 8(1):13

19. Zhang Q, Wu H, Yao P, Zhang W, Gao B, Deng N, Qian H (2018) Sign backpropagation: an on-chip learning algorithm for analog RRAM neuromorphic computing systems. Neural Netw 108:217–223

Chapter 5
Communication Networks for Neuromorphic Systems

Abstract The brain connectivity is generally described at several levels of scale, including synaptic connections that link individual at the microscale, networks connecting neuronal populations at the mesoscale, and brain regions linked by fiber pathways at the macroscale. Since each neuron is connected to many others, high bandwidth is required. Moreover, since the spike times are used to encode information, very low communication latency is also needed. In this chapter, the network used for communication in neuromorphic systems are covered. In particular, the Network-on-Chip fabric is introduced for receiving and transmitting spikes following the Address Event Representation (AER) protocol and the memory accessing method. The interconnect methods for inter-neurons communication is covered in details. Moreover, the interconnect design principle is presented to help understand the overall concept of on-chip and off-chip communication. The remaining parts cover advanced on-chip interconnect technologies, including si-photonic three-dimensional interconnects and fault-tolerant spike routing algorithms.

5.1 Introduction

Early neuromorphic chips were designed by Carver Mead and his students at Caltech [56, 80]. In the first design, Address Event Representation (AER) is used as the interchip communication protocol because a massive number of neurons within the chip cannot be realized at that time. AER multiplexes the firing events from neurons and encodes it as a lower complexity connection. For an N axonal fiber, with one active at a time, AER replaces regular wire with (1+log N) wires.

In recent neuromorphic designs, AER is used as a protocol for on-chip and off-chip communication. Figure 5.1 shows the AER protocol. The neurons in the sender array generate a temporal sequence of digital amplitude events to encode their outputs; a representation conceptually equivalent to a train of action potentials (or a train of spikes). Each neuron has a digital address that is uniquely assigned to it (for instance: 1, 2, and 3). Whenever a neuron signals an event, the encoder circuitry broadcasts that neuron's address on the inter-chip data bus. After an action potential, the neuron enters a refractory period which prohibits its ability to generate

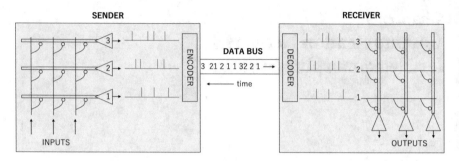

Fig. 5.1 Address-event representation (AER) protocol

new action potentials [74]. Therefore, the inter-spike interval (ISI) at a neuron is longer than the time required to broadcast the neuron's address. As a result, many addresses can be multiplexed on the same bus without major bottleneck. The receiver interprets the address as an event that corresponds to an action potential from the neuron identified by that address.

5.2 Neural Communication

Most biological neurons communicate predominantly via an electrochemical impulse known as an action potential or spike [74]. As most silicon neurons follow the 'point neuron model' [34], the details of dendrite structures are ignored, and we assume all inputs effectively arrive at the neuron. This complex electrochemical pulse is supported by transient sodium, chloride, potassium, and electron fluxes. The size and shape of the spike are invariant, mainly being determined by local instabilities in the cell membrane current balance, so a spike can be viewed as a unit impulse that conveys information solely in the time at which it occurs. It costs the axon energy to transmit an event, but this is provided by a kind of electrochemical 'gain' distributed along the length of the fiber. The net effect is that the axon can be viewed as a lossless dispersion free transmission line, although it has to have a 'rest' just after a pulse has been transmitted, to 'charge itself up' again.

As shown in Fig. 5.1, there is no global clock. In other words, the signals are sent asynchronously; therefore, the spiking time is encapsulated by the event itself, not by the concept of time-step in the digital system synchronization. Once *neuron 1* fires, its spike arrives at the encoder and is encoded to value '1' in the data bus between two modules. After that, there is no spike and no information will be sent via the data bus. Once there are spikes, the neuron's ID is sent to the data bus right after the spike is received at the decoder. In the decoder, the signal from the data bus is read, and the decoder re-issue the spike to mimic the actual communication. Then, the spikes are distributed to the connected post-synaptic neurons via the crossbar circuitry.

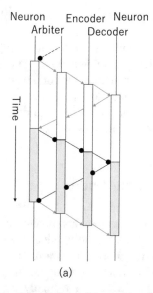
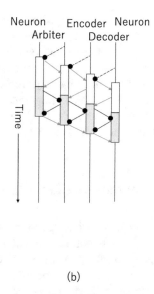

(a)   (b)

Fig. 5.2 Control signal flow starting from a neuron through the arbiter and encoder on the transmitting side, to the decoder and a neuron on the receiving side (left to right). (**a**) Completion of the spike transmission for the originating neuron before the handshaking is completed in the acknowledge phase. (**b**) Pipelining reduces the overall handshaking time by allowing the signals to propagate forward in the set phase without waiting for the acknowledge signal

To communicate between two chips or two neuron clusters, the request and acknowledgment protocols are typically used. At first, a request signal is sent, and later the AER signals are transmitted. When the AER signals arrive and are stored at the receiver, an acknowledge signal is sent back to the sender to confirm the arrival of the spikes. Figure 5.2a illustrates the send/acknowledge handshaking phase between two blocks (cluster of neurons). First, the arbiter allows a pre-synaptic neuron from the sender block to send the signal to the encoder. The encoder then encodes the spike to the AER format and sends the above spikes to the decoder via the data bus. At the decode phase, the AER format signal is unpacked (represented as spikes) and sent to the connected neuron(s). After the spike has been received by the post-synaptic neurons, an acknowledgment signal is sent back to inform the source neuron of a successful reception. Once the acknowledgment signal arrives at the pre-synaptic neuron, the spike is removed, and the arbiter allows other neurons to fire. If there is no acknowledgment, it means the spike is lost during transmission, and a re-transmission is required. Pipelining can be used here to reduce the gap between two communications as described in Fig. 5.2a. The arbiter here allows a new neuron to fire after the encoder receives the signal.

There are two basic types of inter-neural communications: local and global interconnects. While the local interconnect transfers the spikes within a particular area, long-range analog signals help neurons communicate to a different area of the brain. Therefore, a communication network for a given neuromorphic system should consider the local communication and take into account the cluster-to-cluster communication during the design phase.

The communication network in a given neuromorphic system must satisfy the ability to send and receive spikes (action potentials) with low and precise latency. Consequently, unlike early neuromorphic prototypes with hundreds or thousands of neurons, a large-scale system must be explicitly designed to achieve the biological system's real-timeliness. In the next section, we will discuss how to design the interconnect for inter-neural communication in neuromorphic systems.

5.3   Interconnect for Inter-Neural Communication

Even though AER has been the de facto for silicon neurons communication, the protocol itself is not exceptionally scalable. The AER can help reduce the traffic between two group of neurons; however, it is no longer a straightforward solution for systems with a massive amount of neurons. More and more neurons are integrated into clusters with shared encoders and decoders, leading to a highly complex design and long wire length. Therefore, one of the standardized infrastructures for neuromorphic systems is to adopt the on-chip and off-chip interconnects. The major works on neuromorphic systems support a large number of neurocores per system.

The communication between neurons can be done via synchronous [35] or asynchronous manner [30, 30, 62]. While synchronous systems are easy to manage and control, asynchronous systems provide a low power consumption and closer behavior to biological brains which perform asynchronously.

Table 5.1 shows three significant neuromorphic systems: SpiNNaker by the University of Manchester, TrueNorth by IBM, and Loihi by Intel. While SpiNNaker

Table 5.1 Neuromorphic system communication

Architecture	Configuration	Communication
SpiNNaker [35]	Each ARM core perform 1000 neurons' operation. A node consists of 18 ARM cores. 1024 neurons per ARM core. 16-bit for node, 5-bit for core, 11 bit for axon	Nodes are connected using six communication links in triangular lattices folded onto the surface of a toroid. Multi-cast based using CAM
TrueNorth [17]	Each core emulates 256 neuron, 4096 (64 × 64) cores per chip. 18-bit for core distance, 8 bit for axon	Formed in 2D-mesh. Uni-cast based with relative X and Y coordinates. X-first routing
Loihi [30]	128 neuromorphic cores and 3 x86 cores per chip and can be scaled up to 4096 cores. Support up to 16,384 inter-chips communication. Each core implements 1024 neural units. Variable synaptic resolution	Asynchronous 2D Mesh NoC. NoC only supports uni-cast, and the multi-cast is supported by iteration. NoC routing using dimension-order routing algorithm (X-first)

emulates neurons using ARM cores, both TrueNorth and Loihi implement the hardware neurons.

5.3.1 SpiNNaker

The SpiNNaker processor [35] consists of an array of ARM9 cores communicating via packets carried by a custom interconnect fabric. Each packet has 40 or 72 bits and is supported by hardware only. The processor has the ability to transmit over 5 billion packets/s. The 'point neuron model' is also adopted here, leading to the point where all packets (spikes) must effectively arrive at their desired destinations. To represent the connection between neurons, the weights are used to alternate the strength of the connections. The weight can be positive or negative to represent the excitatory and inhibitory connections.

SpiNNaKer adopts the AER protocol as the central idea with its modifications to form the communications between neurons.

Since in AER format, a spike is represented by its time and identity (i.e., number), SpiNNaker uses packet-switched communication and broadcast/multicast routing. Thus, once a spike occurs, the AER signal is sent to the packet-switched communication fabric and delivered to the connected neurons.

By enabling communication via electronic fabrics, the spikes can instantaneously arrive at their destinations, similar to the biological signals in the brain. This allows SpiNNaker to have the freedom to map any neuron to any node (cluster of neurons), and it can virtually form the biological topology regardless of the packet-switched communication topology. The problem of efficient mapping has been extensively investigated to reduce the transmission time between neurons [19].

Since electronic signals can be delivered instantaneously, the *time* information of the spike cannot be preserved. Therefore, it has to be presented in another way. In the biological system, spikes are expected to be at the right place at the right time. However, the electronic system works the other way around. If there is no congestion in the communication infrastructure, the spikes are delivered effectively; however, congestion or failures in the communication system can delay, drop, or misroute spikes. Therefore, there is a need for a scheduling method and fault-tolerant mechanism for neuromorphic systems.

In SpiNNaker, 1000 neurons are simulated in each ARM core using fixed time-multiplexing. Each node of SpiNNaker has 18 ARM processors, which makes a total of 18,000 neurons to be simulated. Therefore, SpiNNaker expects the delivery time (spike window) to be likely at $0.2\,\mu$s/hop to ensure the neurons react to the stimuli in order of ms like biological systems.

At first, the communication between ARM cores of a SpiNNaker chip is handled by a Network-on-Chip. Then, it is converted to off-chip communication using packet-router modules. Six links are merged using a time-division multiplexer to stream together, spikes from the local NoC. The output stream is later split into six output links. The inter-node communication in SpiNNaker is made via packets.

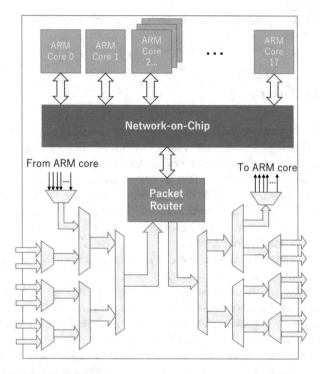

Fig. 5.3 The SpiNNaker node

The packets are generated by cores and transmitted to the local router. The packets are then redirected to the target cores. If the destination neurons are in the same node as the source, the local router sends the packets back to the local cores. If the destination is in other node(s), the local router sends the packets to a neighboring node. Since each node can only connect to six nodes in general, a routing technique is needed to deliver the packets efficiently (Fig. 5.3).

In SpiNNaker, there are four types of packets which are either 40-bit or 72-bit formats: (1) nearest neighbor, (2) point-to-point, (3) neural event multi-cast, and (4) fixed route. While the nearest neighbor packets are used to initialize the system, the flood-fill communication (broadcast) and debugging, and point-to-point packets, allow more detailed communications. Among those packets, we focus on the *neural event multi-cast* as it represents the interneural communication. Here, the AER of SpiNNaker can be summarized in Fig. 5.4 where it uses 16 bits for node ID, 5 bit for the core ID (ARM core), and 11 bits for the neuron ID (neuron within ARM core). The routing method for AER follows the Content-Access-Memory (CAM) method, where it first looks-up the CAM to find the address on the output RAM. The word of output RAM of 6 + 18 encodes the output directly in one-hot format (first 6 bit are for the inter-chip link and last 18 bit for the internal core).

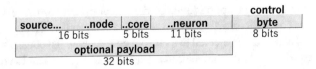

Fig. 5.4 The SpiNNaker AER

5.3.2   TrueNorth

The TrueNorth [17] is a neuromorphic chip designed by IBM in 2014 [17]. The chip consists of 4096 cores connected via Network-on-Chip. Each core has 256 simulated neurons working in time multiplexed method. Each neuron has 256 programmable synapses that emulate the strength between two neurons. Unlike SpiNNaker, which uses ARM core to emulate neurons, TrueNorth implements the 'Leaky-Integrate-and-Fire' model. The number of connections is also limited to 256 per neuron; however, it achieves much lower power than the large-scale system of SpiNNaker. Unlike SpiNNaker, TrueNorth does not use the global address of the node/cluster. Instead, it uses the relative distance in X and Y (dx and dy) to achieve communication. TrueNorth uses a 9-bit sign integer for both coordinates, which allows packets to be sent in a range of $[-256, 255]$ from the source.

The TrueNorth uses the X-first routing algorithm where the relative distance in X-coordinate (dx) value is increased or decreased first until it gets to zero. Then, the relative distance in Y-coordinate (dy) value is changed. In each hop, unless both dx and dy are zero (which means the packet arrives to its destination), the packet is routed based on the change of dx and dy values. For instance, if the dx is negative and increased in the routing unit, it is routed to east.

The content of the spike data besides dx and dy are delivery tick (4-bit), destination axon index (8-bit) and debug bits (2-bit). While the delivery tick is used to ensure the proper timing of spike, the 8-bit axon index represents the actual connections. In TrueNorth, the synapse SRAM is 256×410 for 256 neurons. Here the synapse resolution is one bit per connection.

The routing is TrueNorth is based on unicast with an X-first routing algorithm. Since the spikes are sparse, deadlock and livelock are not a concern.

5.3.3   Loihi

In 2018, Intel released its $60\,\text{mm}^2$ neuromorphic chip, named Loihi, fabricated in 14 nm process. The Loihi chip consists of 3×86 cores and 128 neuromorphic cores connected via an asynchronous NoC and can be scaled up to 4096 cores. Each core can implement up to 1024 neurons with variable weight length (1–9 bits), which also varies the number of synapses per core (1 million down to 114k). The NoC supports no native multicast, and the core has to send to a list of destinations iteratively.

Although there is no information about the actual packet format of Loihi, there is consensus that an AER-like format is used for sending spike. Once a pre-synaptic neuron fires, its spikes are projected to a fan-out core-to-core list consisting of the ID list of the core with post-synaptic neurons. Then, the spikes are sent to all connected cores using unicast via the NoC. Once it is received at the destination cores, it performs a multicast transmission to all post-synaptic neurons.

5.4 Interconnect Design Principles

The previous section has surveyed the three notable system designs for neuromorphic systems. Since we focus more on neural communication in this section, we summarize the major principles for designing the neuromorphic system interconnects.

- *Network topology:* before providing any features in the system, deciding the topology of the system is a critical issue. Different topologies have different advantages and drawbacks. As we survey in Sect. 5.3, SpiNNaker uses a folded triangle lattice type of topology while other works simply use 2D-Mesh. The topology of SpiNNaker can fit the supercomputer model; however, 2D Mesh is a simple and scalable topology to use when it comes to neuromorphic chips.
- *Classification of the communications:* Beside the *inter-neural communication* that allow spikes to travel from pre-synaptic to post-synaptic neurons, the *synchronization, data transaction* and *debugging* are also needed. Although most of the reports do not focus on how to program the neuromorphic system, the amount of weights and parameters need to be programmed are considerable. On the other hand, synchronizing the system (i.e., increase the time-step) is also important. Here, flooding or gossiping is the common approach where the core exchanges information with its neighbors to allow a system synchronize.
- *Network support* for the communication is another issue. The communications can be classified into three main categories: *uni-cast, multi-cast* and *broad-cast*. While SpiNNaker support *multi-cast*, both TrueNorth and Loihi NoCs only support *uni-cast* which require the source node to send multiple packets to multiple destination. On the other hand, flooding and gossiping communication performing the function of the *broad-cast*, and *uni-cast* based approach is more viable.
- *Time constraints* of the neuromorphic system are mostly on the inference time where the spikes must arrive at their destination within a certain window. While SpiNNaker relies on the interconnect infrastructure to deliver the packets through the whole system within a millisecond to obtain *biological real time*, designers can expect to accelerate the execution time by reducing the transmission time of the network. While the inter-neural communication has two type of connections: *local* and *long-range*, the *long-range* connections are the critical path to ensure real timeliness and accelerated performance. Although it is not recommended,

spike dropping can be utilized to have the system perform under a certain level of accuracy loss. The critical traveling path must be addressed properly. Here, the mapping problem of the neuromorphic system can help solve the problem.

- *Mapping* is another problem that should be carefully addressed. The large-scale system can select different neurons in different clusters/nodes to map a neural functions. Recent works [19] on mapping have tried to address by reducing the communication time or execution time.
- *Fault-tolerance* support is another feature for the neuromorphic systems. Although neuromorphic systems are naturally resilient to some faults, faulty interconnect can lead to missing spikes or misrouted spikes and change the system's behavior.

5.4.1 OSI Model for Network-on-Chip

NoCs are usually packet-switched communication networks derived from the general computer network domain. The OSI (Open Systems Interconnection Model) reference model for NoC is depicted in Fig. 5.5. This acts as the framework to build the NoC. The standard OSI models are classified into seven layers as shown in Fig. 5.5.

5.4.1.1 Application and Presentation Layers

The application and presentation layers are where the data-flow graph can be represented. Each task in the data-flow graph has a different workload, and the tasks are connected via messages or transactions. In a neuromorphic system, such

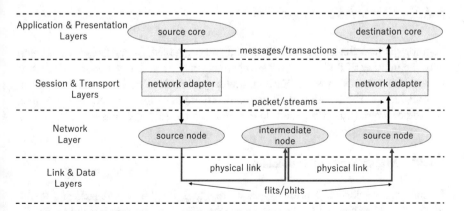

Fig. 5.5 OSI reference model for Network-on-Chip

data types could be spikes, data transactions (read/write local memory), or control signals.

5.4.1.2   Session and Transport Layers

The messages and transactions will be divided into packets or streams and delivered to the network interface to communicate. The destination, routing path, and routing instruction can be encapsulated into the packet in this layer. Quality of services such as best-effort or guaranteed services is also delivered at this layer.

5.4.1.3   Network Layer

After encapsulating the packets, the network interfaces send them to the network in the network layer. Finally, the network delivers the packets from the source network interface (NI) to the destination NI via intermediate routers. We could have different switching techniques such as circuit switching or packet-switching (store-and-forward, virtual-cut-through, or wormhole) in this layer.

5.4.1.4   Link and Data Layers

These layers are where the packet is split into a flow control unit (flit). Each flit is received, routed, and sent by sub-modules of routers. The flow control needs to be established to ensure the synchronization between routers. Moreover, protecting the data integrity using error correction codes is also necessary.

5.4.2   Network Topologies

Topology is how the network is structured and organized. In particular, it has to do with the number of nodes or processing elements (PE) in the system, the design of routers, the connecting wires between routers and router-PE, and how a packet can be transferred. The suitable NoC topology, communication mechanism, PEs type, and router architecture are generally decided through extensive design exploration.

5.4.2.1   Major Types of Topologies

In interconnect designs, various types of topology can be considered. In general, they can be classified either as *regular* or *irregular*. Regular topologies are usually designed for a particular application where it fits exactly the communication requirements of the system. However, Irregular topologies are typically unscalable and cannot fully exploit the parallelism of NoCs. In the scope of this neuromorphic book, we only considered the regular topologies.

Several types of regular topologies have been proposed, and they include: *mesh*, *torus, folded torus, concentrated mesh, binary tree, butterfly tree*, and *clos*. Each topology has different advantages with regards to distance between core, number of cores attached to router, routing complexity, and layout optimum. Among these topologies, *mesh* topology can be considered as the most common topology, thanks to its simplicity in design, routing, and layout. The following sections will cover the significant mesh topologies.

5.4.2.2 2D Mesh Topology

The 2D Mesh topology is depicted schematically in Fig. 5.6. A 2D Mesh NoC consists of $M \times N$ mesh of routers (or switches) interconnecting Processing Element (PE) placed near the router. Each router, except those at the edge, can connect to four neighboring routers and one PE. The number of routers is equivalent to the number of PEs in the system. The communication channel between routers or between routers and PE consists of two unidirectional links.

In Mesh topology, there is an assumption that the links are in similar lengths; therefore, this allows simple physical designs. This also usually implies that a similar architecture is shared between PEs. As a result, one tile (a combination

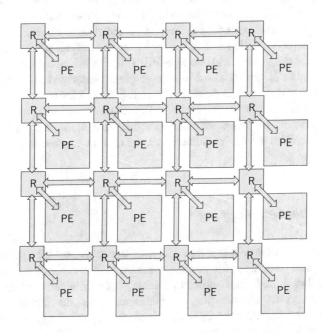

Fig. 5.6 2D Mesh topology

of one router, NI, and one PE) is generally designed first as a macroblock and then replicated in the physical design phase. This can accelerate the design time and allow highly scalable systems. A multiple-chip system can be established in neuromorphic systems by extending the NoC in a 2D plane, and no routing adjustment needed.

The communication in 2D Mesh is established by packaging data in flits (flow control units), which is the amount of data a link can send at once. Flits can be attached to create a packet that can be sent serially. Flits are sent from the PE via its network interface, travels through the routers in the network using a routing algorithm, and delivered at the destination PE.

Several routing algorithms have been created for 2D-Mesh, such as XY, West-first, North-last, or Negative-first. These routing algorithms are tableless as no routing table is needed. The relation between the current node, destination, and the incoming port is used to perform such routing. Several adaptive routing algorithms can deal with congestion or faults in the network. The details about the communication architecture, such as switching technique, packet routing, and quality of service, will be further discussed in Sect. 5.4.4.

5.4.2.3   3D Mesh Topology

The three-dimensional (3D) stacking technology is an emerging technology that helps keep the momentum of Moore's law which allows double integration of transistors every two years. As dealing with cross-layer communication is an issue, inheriting 2D Mesh and expanding to the vertical direction is a natural solution.

Figure 5.7 depicts the 3D Mesh topology where each router can be connected to up to six neighboring routers and one local PE. This also maintains the same ratio of router and PE as in 2D Mesh. The design principle is unchanged in 3D Mesh topology unless a new direction is created. Through-Silicon-Vias are used in stacking 3D-ICs to allow inter-layer communication. Recent development in Monolithic 3D-ICs can support multiple layers with smaller sizes. The communication of 3D-Mesh can be inherited from the 2D-Mesh; however, the routing now needs to consider the Z-direction.

5.4.3   Application Mapping

After defining the network's topology, the next step is to map the application into the network. The mapping problem deals with how to map the application tasks into the target architecture. Here, bandwidth, latency, and architecture parameters, such as area cost and power consumption, are considered during the mapping process. In general, the communicating blocks should be placed in a close region if they require a considerable bandwidth or low latency.

Fig. 5.7 3D Mesh topology

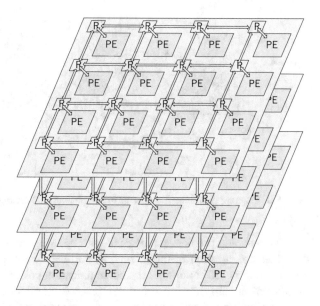

The mapping algorithms usually start with the data-flow graph and the constraints of the applications (i.e., latency or execution time). In the scope of this book, this section focuses on mapping neuromorphic application into a NoC-based system. Therefore, the data-flow graph can be randomly connected networks such as liquid state machines, reservoir computing, or multiple layer networks (feed-forward fully connected or convolution neural network). Depending on the mapping application, other constraints, such as the maximum bandwidth of the packet or real-time requirement, can be considered. The mapping issue for NoC is considered as NP-hard; therefore, complexity reduction is needed as a heuristic method is not suitable.

Mapping neuromorphic systems conventionally consists of two phases: (1) *partitioning* and (2) *placing*. The partitioning is to determine the cluster of each node. In general, neuromorphic systems are divided into clusters of neurons. The number of neurons per cluster is usually limited (i.e., TrueNorth dedicates 256 neurons per cluster). Therefore, the partitioning process will decide which cluster the neurons belong to. The general idea is to place the neurons that share the communication or heavily connected in the same group. This ensures that fewer packets are sent via the NoC. Figure 5.8 describes the process of *partitioning* nine neurons (in circle) into three clusters. As shown in the figure, the strongly connected neurons are placed in the same cluster to avoid congestion in the network.

After the partitioning process, the mapping process is now converted from neuron-to-neuron communication to node-to-node communication. Figure 5.8 illustrate the placing process as each cluster is assigned to a PE. Because the bandwidth between cluster 1 and the other two clusters is high, the other clusters are placed as neighbors of cluster 1. Cluster 2 and cluster 3 are placed in the distance as the required traffic is not high.

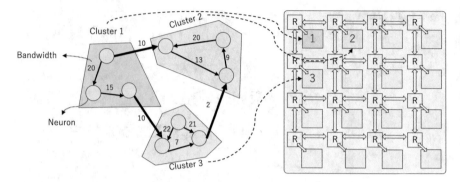

Fig. 5.8 Mapping neuromorphic systems in two phases: *partitioning* and *placing*

The mapping problem can be dealt with by using Integer a Linear Programming or other optimization methods, such as Particle Swarm Optimization [20]. While the method in [20] optimizes the communication cost, work in [84] optimizes the traffic in on-chip network based neuromorphic systems. In SpiNNaker [34], the linear mapping method is used as neurons are uniformly allocated into NoC nodes in their index order. This method is not optimal; however, this reduces the mapping time significantly as it is not easy to map at a large scale. In [44], mapping deep neural network for SpiNNaker is also presented. In Chap. 9, we show the mapping solution using a genetic algorithm. In summary, a mapping problem is generally solved with an optimization technique to reduce the traveling distance of spikes.

5.4.4   Communication Architecture

After having the topology and a proper mapping solution, designing the communication architecture to support the interconnects between PEs is the final task. This section covers the critical parts of the communication architecture, including switching technique, packet routing, quality of service, and the design of router and link.

5.4.4.1   Switching Technique

The switching technique determines how an input port of a router is connected to an output port of a nearby router. First, data are usually sent in a message and split into packets. Later, a packet will be divided into a flow control unit (flit) and then physical units (phits).

Two commonly used techniques are packet switching and circuit switching. In packet switching, there are three standard methods: store-and-forward, virtual-cut-

Table 5.2 Comparison of switching techniques

Method	Performance	Buffer size	Design complexity	Flexibility	Transmitting latency
Circuit switching	High for long messages	1 flit	Low	None	Very low
Store-and-forward	High for short messages	Packet	High	Low	Low
Virtual-cut-through	High	Packet	High	Low	Low
Wormhole	High	A few flits	Low	Medium	Low (if not blocked)

through, and wormhole. The comparison between these techniques is presented in Table 5.2.

The circuit switching technique is operated by dedicating the whole channel for a requested communication between two nodes. The path will be reserved for sending the messages and will be released when the transmission is done. During the reservation time, no other packets whose path overlaps the reversed path can be sent. As a result, circuit switching is suitable for long messages. With a short message, the time to reserve and release the path can be problematic. Also, there is no flexibility in routing algorithms because the path is defined. There are three practical approaches in packet switching: store-and-forward, virtual-cut-through, and wormhole. In store-and-forward (SAF), each router can store a whole packet. Packets can only be sent to the next router once it is ultimately delivered at the current router. On the other hand, virtual-cut-through allows sending the packet as long as the next router can host the new packet. This can reduce the latency because the router does not need to wait for the full packet to be received before sending it. Since both methods need to store the packet in the buffer, the buffer size is significant, making the design more complex.

The wormhole switching can be used to reduce the area cost as it only stores a few flits and will forward the flit as long as the next router allows. To do so, the handshaking mechanism between the router needs to ensure no overflow in the receiving buffer. Thus, the wormhole technique allows a smaller buffer and less complex design. Also, it will enable more complexity in the routing. The major drawback of the wormhole is high latency if the packet is blocked and a potential locking situation.

To solve the blocking situation, *virtual channel* is proposed. Instead of a single buffer, several buffers are dedicated to virtual channels. Time-division-multiplexing (TDM) is used for switching between channels. This allows a blocked packet to be passed by a non-blocking packet.

5.4.4.2 Packet Routing

Another key aspect of designing NoC is how to route a packet. The topology closely constrains the routing. In the scope of this book, we only mention the routing for mesh topology. Other topologies can have different approaches to routing. The major approaches for routing are:

- By destination address or coordinates (i.e., (0,1,2) in 3D-NoC). This one can be translated to the differences in coordinates (i.e., (0,2,2) to (1, 2, 0) by using these field (+1, 0, −2) which is updated every time the packet moves).
- By a group identifier (i.e., group A or layer 2). This method is usually used for multicasting.
- By predetermined path (i.e., go by the path: North, West, West, Local). This is usually used for deterministic routing.

There are three types of routing algorithms:

- Deterministic routing: the routing path is fixed. Notably, the routing path is usually decided at the source node and cannot be changed.
- Adaptive routing: the direction of movement is decided on each router.
- Stochastic: the path choice is varied between different routing paths at different probability.

The routing algorithm could also be *minimal* or *non-minimal* ones. The minimal routing always makes sure the number of hops a packet needs to travel is the smallest one, while non-minimal routing allows variations.

In NoC routing, *deadlock* and *livelock* are two potential problems. *Deadlock* usually occurs when a cyclic dependence graph is formed between packets/router. Since a packet needs other packets to be routed so as to release the routing resource, the cyclic dependence graph forever waits for the packet. To solve this issue, detecting a cyclic dependence graph can be used, and a packet will be dropped to solve the issue. Prevention can be obtained by partial routing, which prohibits some turns which do not form the cycle. For instance, 2D-Mesh XY routing always routes X first. Therefore, direction such as up-right or down-left is not allowed. The other routing algorithm that prohibits cyclic routing paths is west-first, north-last, or negative first.

Livelock happens when the packet is routable; however, it cannot reach the destination due to a blocking situation (i.e., meet a deadlock or faulty path). To prevent livelock, *non-minimal* routing is prohibited. If there is non-minimal routing in the system, detecting the livelock packet and dropping it is necessary.

Another approach for routing is to use a Look-up-Table (LUT). LUT consists of the routing direction for each destination. Instead of calculating the direction, the router only needs to find the direction in the table. This method is usually used for group routing as there is no actual address for a group.

5.4.4.3   Flow Control

Another aspect of communication architecture is how to control the flow of flits between routers or routers and PEs. This chapter only covers the buffer flow control as the conventional NoC design for neuromorphic systems is the packet-switching wormhole.

There are three common flow control methods:

- *ACK/NACK*: This protocol is a handshaking protocol between two terminals (router-router or router-PE). When the sender tries to put data on the link, a valid signal is also sent to the receiver. If the receiver is ready to take the data and store it in its buffer, an ACK signal is sent. Otherwise, a NACK is sent to let the sender hold the data. If the sender receives an ACK, it will send the data as the receiver will take it.
- *STALL/GO*: This protocol allows the receiver to inform the sender whether it is ready to receive data. The sender will send a request to the receiver to send a flit. Once the 'GO' signal is one, the sender will send the flit.

    CREDIT: This protocol informs the credit of the input buffer, which is usually the index of the reading pointer. Based on the credit, the sender can understand the status of the buffer in the receiver.

5.4.4.4   Quality of Service

In NoC architecture, some applications need to tune the transmission time of a packet or flit. For instance, spikes need to be routed in the neuromorphic system as fast as possible to allow fast execution time. Therefore, NoC also supports Quality of Service, such as best effort (BE) or guarantee service (GS).

Best effort (BE) usually does not reserve any routing resource. Therefore, the delay cannot be guaranteed. If the NoC allows dropping packets, there is a chance the packet will not be delivered.

NoC also can guarantee the delivery with the pre-allocated resource. For instance, packet-switching with time-division-multiple-access (TDMA) and pure circuit switching can be used to ensure the packet's time and delivery. This method is usually used for real-time applications.

5.4.4.5   Router Design

In this section, the design of a generic router will be covered. The router design is related to the NoC parameters presented above, including topology, flow control, and the switching technique. The case study in the next section will show a detailed router with fault-tolerance and multicast routing. There are several design parameters for the router:

- *number of ports:* Depending on the topology, and the location of the router, the number of ports will be different. For instance, the 2D-Mesh router usually has five ports, while the 3D-Mesh router has seven ports. In 2D-Mesh, the border and corner router have four and three ports, respectively.
- *phit sizes:* The number of parallel wires of a router link. Typically, the phit sizes are equal in all ports. In 3D-Mesh, the inter-layer links are sometimes designed in a time-division multiplex mode which uses fewer wires than usual.
- *buffer design:* The placement, size, and queuing method of the buffer are also important. For instance, the case study we show later has no output buffer, but uses FIFO 4-flit for the input buffer.

Figure 5.9 depicts the generic architecture of a router with several input ports and output ports. In each input port, there is an input buffer and a routing unit. The routing unit decides which direction the packet needs to go by using a routing algorithm. Then, the routing unit in each input port sends a request signal to an output port allocator. This module will decide which input port is granted access to the crossbar. If an input port is granted access, it sends the flit through the crossbar to a proper output port. Flits are stored in the output buffer at the output port to send to the next router or the attached PE.

Input Port Input port is usually made up of two major modules: input buffer and routing unit (see Fig. 5.10). The input buffer can be an SRAM or normal registers. An input buffer can have multiple virtual channels, and each channel can store several flits. If there are several channels, the routing unit needs to decide which channel can be used to route. Once the channel is selected, the routing of the packet

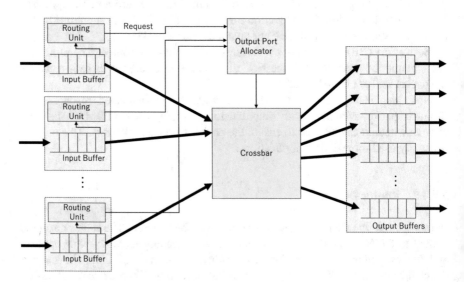

Fig. 5.9 The generic architecture of a router

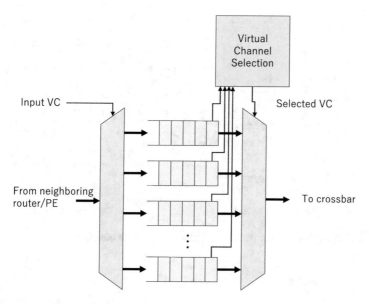

Fig. 5.10 Network Interface architecture

in the channel is loaded. The routing unit will decide the output port based on the routing algorithm.

Output Port Allocator Receives a request from the input ports and decides which input port can be granted access to the crossbar. If there is no conflict (i.e., all input ports request different output ports), it will grant all requesting input ports. If there are conflicts, on of the following scheduling policy is used:

- *Round-robin:* Each input port is assigned a slot in a ring counter token. If there is no other request before an input port, the input port is granted. Otherwise, it needs to wait. Once an input port is granted, the ring token is turned, which periodically gives each input port the highest priority.
- *LRU:* The least recently used first method increases the priority of an input port relative to the time it was granted. Once an input port is granted, its priority drops to the lowest and will be gradually increased.
- *Fixed priority:* Each input port is assigned a fixed priority. This is the easiest method to implement; however, it can create unbalanced priority which leads to some paths always sending packets first.

After an input port has been granted, it can send the flit to the crossbar or switch fabric as depicted in Fig. 5.11. There are two methods to design the crossbar: (1) pass transistors and (2) multiplexers. The pass transistor is intended to be an actual crossbar, and the transistor will shorten the connection from an input to an output port if the connection is granted. The multiplexer design uses a multiplexer for each output port and selects one input signal to drive.

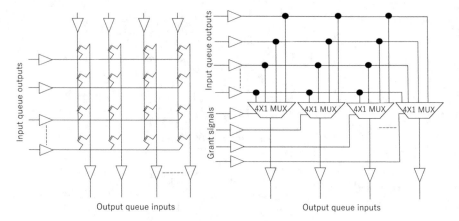

Fig. 5.11 The architecture of a crossbar: (**a**) pass transistors, (**b**) multiplexers

5.4.4.6   Link Design

The link between two routers or router-PE is a group of wires. The following design issues for a link should be considered:

- *Wire delay*: The distance between two routers can be long due to the large distance between PEs. This can lead to high delay wires, and violate timing constraints. CAD tools can automatically insert buffers (or a couple of inverters) to maintain the strength of the signal; however, if the delay is significant, inserting registers between two routers can help reduce the delay. Please note that the flow control can be changed when registers are inserted.
- *Vertical links*: In 3D-NoCs, vertical links can be Through-Silicon-Vias (TSV) or inter-layer via in Monolithic 3D. Since TSVs are not well supported by the current CAD tools, placing TSV needs to be carefully considered. Thermal variation of TSV can also create stress on other devices; therefore, Keep-out-Zones around TSVs are necessary.

5.5   Advanced Interconnects Multicore Neuromorphic Systems

Emerging applications are getting increasingly complex, requiring good architecture to ensure sufficient bandwidth for any transaction between memories and cores, and communication between different cores on the same chip. Because of these and other factors, conventional 2D-NoC interconnects have become an unsuitable candidate for future large-scale neuromorphic systems that are expected to accommodate hundreds of neural clusters and millions of neurons. This section discusses advanced technologies for interconnects toward neuromorphic systems.

In 2D-NoC, if a given packet traverses many hops to reach its destination, the communication latency will be long, and consequently, the throughput will be lower. In other words, an extensive network diameter has a negative impact on the worst-case routing latency in the system.

The need for optimizing 2D-NoC based architecture have become more necessary, and many researchers have been able to achieve this goal with various approaches, such as developing fast routers [6, 9, 45, 46, 66] or designing new high-throughput, and low latency network topologies [28, 47, 69]. One of these proposed solutions was porting the 2D-NoC architecture to the third dimension [72]. In the past few years, 3D-ICs have attracted a lot of attention as a potential solution to resolve the interconnect bottlenecks. A 3D chip is a stack of multiple device layers with direct vertical interconnects tunneling through them [29, 65].

So far, results achieved in this area have shown that 3D-ICs can achieve higher packing density due to the addition of a third dimension to the conventional two-dimensional layout; thanks to the reduced average interconnect length, 3D-ICs can achieve higher performance. Apart from the vital benefit of reduction in total wiring, a lower interconnect power consumption can also be obtained [40, 81], not forgetting that circuitry is more immune to noise with 3D-ICs [72]. This may offer an opportunity to continue performance improvements using CMOS technology with smaller form factors, higher integration densities, and supporting the realization of mixed-technology chips [24]. As Topol [81] stated, 3D-IC can improve the performance even in the absence of scalability.

3D NoC architecture is expected to meet scaling demands for future multicore and many-core SoCs, exploiting short vertical links between adjacent layers that can clearly enhance system performance. This combination is expected to provide a new horizon of NoC and IC designs in general.

One of the essential design steps that should be taken into consideration while designing a 3D-NoC is to implement an efficient router since it is the backbone of any NoC architecture. A router's performance depends on factors and techniques, such as the traffic pattern, router pipeline design, and the network topology. As stated, among these three factors, we have less control over the traffic patterns compared to the topology and the pipeline design. Following this logic and assuming the topology choice was already taken, one of the most crucial router enhancements that can be done is to improve the pipeline design. Reducing the pipeline delay via pipelining optimization decreases the per-hop delay, and the whole network latency will be reduced. On the other hand, the pipeline design is strongly associated with the adopted routing algorithm. Routing is the process of determining the path that a flit should take between a source and a destination node.

When routing, minimal routing schemes are shorter and require less complex hardware, but allowing non-minimal routing increases path diversity and decreases network congestion. Routing algorithms can be adaptive, where routing decisions are made based on the network congestion status and other information about network links or buffer occupancy of the neighboring nodes. Routing algorithms can also be deterministic.

There are a large number of sophisticated adaptive routing algorithms. However, they require more hardware and are challenging to implement. That's why deterministic routing schemes have been adopted for 3D-NoC designs. One of the well-used routing schemes used in 3D-NoCs is the Dimension Order Routing (DOR) XYZ algorithm. XYZ is a simple scheme, easy to implement and free of deadlock and life-lock. But on the other hand, it suffers from a non-efficient pipeline stage usage. This can introduce an additional packet latency which has an essential effect on the router delay and eventually on the system's overall performance. Enhancing this algorithm while keeping its simplicity may improve the system performance by reducing the packet delay.

A 2D-NoC, named OASIS NoC, was presented in [2, 3, 63, 64]. Although this architecture has its advantages over the shared-bus based systems, it also has several limitations such as high power consumption, high-cost communication, and low throughput.

The presented 3D-OASIS-NoC (3D-ONoC) is based on a so-called Look-ahead-XYZ (LA-XYZ) routing algorithm [5]. This algorithm improves the router pipeline design by parallelizing some stages while taking advantage of the simplicity of conventional XYZ. As a result, this routing scheme enhances router performance, thereby achieving a low-latency design.

5.5.1 Three Dimensional On-chip Interconnect

As we stated in Chap. 1, the number of transistors on a chip have increased over the past few decades, which made shrinking chip size while maintaining high performance possible. This technology scaling has allowed Systems-on-Chip (SoCs) to grow continuously in component count and complexity, which significantly led to some very challenging problems, such us power dissipation, and resource management [38, 53].

Since moving to deep submicron technology poses significant design and manufacturing problems, the 3D integration becomes an attractive option to meet power and performance demands. By stacking dies or wafers, we can reduce wire lengths. As a result, performance is increased, and power consumption is reduced. Thus, the on-chip interconnection network plays a more critical role in determining the performance and power consumption of the entire chip [50].

5.5.1.1 3D-NoC Versus 2D-NoC

The 3D-NoC is a widely studied research topic, and many related works have been conducted in the past. Few of them focused on the benefits of the 3D-NoC architecture over the traditional 2D-NoC design. Feero [32] showed that 3D-NoC could reduce latency and the energy per packet by decreasing the number of hops by 40%, which is an essential factor to evaluate the system performance [32]. Pavlidis [71] analyzed the zero-load latency and power consumption, and demonstrated that a decrease of 62% and 58% in power consumption can be achieved with 3D-NoC when compared to a traditional 2D-NoC topology for a network size of $N = 128$ and $N = 256$ nodes, respectively, where N is the number of cores connected in the network. Power consumption reduction can be related to decrease in the number of hops since a flit has fewer hops to traverse when moving from source to destination, and that includes less buffer access, less switch arbitration, and less link and crossbar traversal. All of these factors will eventually lead to a decrease of the power consumption.

Other part of previous works focused on router architecture. For example, Li [54] has modified the conventional 7×7 3D router using a shared bus as a communication interface between the different layers of the router, to create a *3D NoC-Bus Hybrid* router. This kind of modification reduces the number of ports in each router from 7 to 6. However, flits that want to traverse from one layer to another has to compete for access to the shared bus since it's the only inter-layer communication interface. This may lead to undesirable performance degradation, especially under heavy inter-layer traffic.

Yan [88] also proposed another architecture for the 3D-router by implementing all the vertical links into a single 3D-crossbar. In this case, the router has only five ports since we do not need any more additional ports for the vertical connections. This technique reduces the interlayer distance and makes the travel between the different layers in one single hop possible. But this router also engenders a high router cost. Besides, the implementation complexity of such router cannot be acceptable for some simple application that does not need such a complex router.

Considering all these facts, we adopted the conventional 7×7 3D-router, as it is the lowest cost among the other architectures and also the simplest to implement showing several properties like regularity, concurrent data transmission, and controlled electrical parameters [36, 39]. All the benefits are acquired while making sure that this low cost and simple implementation does not affect the performance of the system.

5.5.1.2   Routing Algorithms

Many routing algorithms have been proposed for multicore system on chip (MCSoC) systems, but most of them focus only on 2D-network topologies. Also, among all the studies conducted for 3D-NoC, few of them concentrate on routing algorithms. For instance, Ramanujam [77] presented an oblivious routing algorithm called randomized partially minimal (RPM) that aims to load balance the traffic along with the network, improving then the worst-case scenario. RPM sends packets to a random layer first, then routes them along their X and Y dimensions using XY or YX routing with equal probability. Finally, packets are sent to their final destination along the Z dimension.

In a pretty similar technique, Chao [26] addressed the thermal power problem which is one of the most important issues in 3D-NoC designs. Starting from the fact that upper the layer in a network detains the highest thermal power in the design, they proposed a thermal aware downward routing scheme that sends first the traffic to a lower layer, routes along the X and Y dimension before sending the packets back up to their destination layer. This technique avoids communication in higher layers, where the thermal power is more critical than the lower ones, and then may reduce the overall thermal power in the design. Thus, ensuring thermal safety while guaranteeing less performance impact from temperature regulation.

Both of these routing algorithms have their advantages in terms of load balancing and thermal power reduction. But the routing used is not minimal, which directly affects the number of hops. By adopting a non-minimal routing, the packet delay may increase in the system, especially when we talk about many connected nodes.

To ensure a minimal path for flits when traversing the network, majority of the 3D-NoC systems have been using the conventional minimal Dimension Order Routing (DOR) XYZ routing scheme. Others introduced a routing scheme based upon XYZ, such as the case of *Tyagi* in [83] who extended a previous routing algorithm [61] called *BDOR* designated for 2D-NoC. *BDOR* forwards packets in one of two routes (XY- or YX-orders), depending on the relative position of a source-destination pair, and that aims to improve the balance of paths along with the network also when taking into account the destination.

The XYZ routing scheme and all the routing algorithms based upon it, are described as vertically balanced routing algorithms that have the best performance, since they are simple to implement, it is free of deadlock and life-lock, and also because packet ordering is not required [26, 31, 52]. On the other hand, it cannot always make the best use of each pipeline stage because the switch allocation stage (SA) is always dependent on the previous routing calculation (RC). This dependency can be explained by the fact that the SA stage needs information about the desired output-port calculated from the RC stage, where the incoming flits should go through to pass to the next neighboring node. To solve the problem in 2D-NoC systems, an intelligent pipeline design can be adopted with the help of some advanced techniques like look-ahead routing [83]. This kind of routing has been used to reduce the pipeline stages in the router by parallelizing some of these stages, then decreasing the router delay, and then enhancing the system performance. Look-ahead routing has been used with 2D-NoC, but it hasn't been adopted for 3D Network-on-Chip architectures before.

A second problem that can be seen with a lot of conventional routers using XYZ-based routing schemes is in case of no-load traffic. When the input buffer is empty, the flit entering the router is first stored in the input buffer before advancing to the next RC stage, even if there is no flit under process in the following stages. This unnecessary stall will increase the packet latency in the router, and its associated power consumption, adding a performance overhead to the whole system even in a light traffic case where the system is supposed to have a close-to-optimal performance since no congestion may increase the latency. To address this problem, a technique called no-load bypass is used [86]. This technique allows the flit to advance to the RC stage, where the buffer is empty, then overlap the unnecessary buffer writing stage (BW). This technique decreases the router delay.

In [1, 2, 11], the design and performance of so called 3D-OASIS-NoC was evaluated using a simple application that randomly generates flits and sends them through the network. But, an actual application could not be assessed due to the absence of some components in the design. For that reason, a network interface has been added to 3D-ONoC, the optimized version of 3D-ONoC, enables a system to be evaluated with real selected target applications (JPEG encoder and Matrix Multiplication).

5.5.1.3   3D-NoC Router Architecture Design

The router is considered the backbone element in the whole 3D-ONoC design. The 3D-ONoC router architecture is based upon the 5×5 2D-ONoC router where each switch has a maximum number of 7-input by 7-output port, where four ports are dedicated to connecting to the neighboring routers in north, east, south and west direction using the intra-layer links. One port is used to connect the router to the local computation tile, where the packet can be injected into or ejected from the network. The remaining two ports are added to connect the switch to the upper and lower layers to ensure inter-layer communication.

As we previously stated, the number of ports depends on the position of the router in the design, since we have to eliminate any unused links that have no connections with other routers to reduce power consumption. Figure 5.12 represents 3D-ONoC

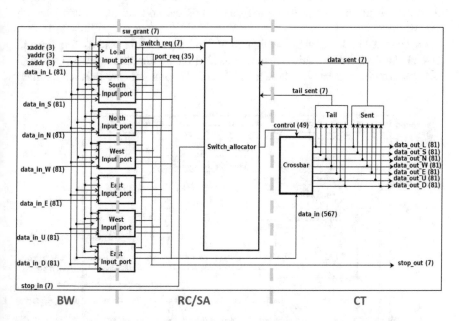

Fig. 5.12 3D-ONoC pipeline stages: Buffer writing (BW), routing calculation and switch allocation (RC/SA) and crossbar traversal stage (ct)

router architecture and the three main pipeline stages can define the routing process at each router: Buffer writing (BW), Routing Calculation and Switch Allocation (RC/SA), and the Crossbar Traversal stages (CT).

5.5.1.4 Input-Port Module Design

Starting with the *Input-port* module represented in Fig. 5.13, each one of the seven modules is composed of two main elements: *Input buffer* and the *Route* module. Incoming 81-bit flits *data-in* from different neighboring switches, or from the connected computation tile, are first stored in the *Input buffer* while waiting to be processed. This step is considered the first pipeline stage of the flit's life-cycle (BW). The arbitration between different flits is managed using the FIFO queue technique. Each input buffer has by default four as depth, which means that it can host up to four 81 bits of flits. The buffers occupy a significant portion of the router area but can also imply an increase in overall performance.

After being stored, the flit is fetched from the *FIFO* buffer and advanced to the next pipeline stage (RC/SA). The destination addresses (*xdest*, *ydest* and *zdest*) are then decoded in order to extract the information about the destination address in addition to the *Next-Port* pre-calculated in the previous upstream node. Those values are then sent to the *Route* circuit where the La-XYZ routing scheme is executed to determine the *New-next-Port* direction for the next downstream node. At the same time, the *Next-Port* identifier is also used to generate the request for the *Switch-Allocator* asking for a grant to use the selected output port via *sw-req* and *port req* signals.

As we stated in the previous section, 3D-ONoC uses a lookahead routing scheme *LA-XYZ* for fast routing. This scheme is based upon the dimension order (DOR) X-Y-Z static routing algorithm, where the X, Y, and Z coordinates are satisfied in order.

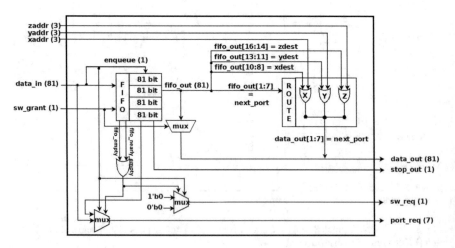

Fig. 5.13 Input-port module architecture

X-Y-Z routing is presented as the vertically balanced routing algorithm, which has the best performance since it's simple to implement. It is free of deadlock and live-lock, and also because packet ordering is not required. In addition to that, each flit additionally carries one hot encoded *Next-Port* identifier used by the downstream router. Since *LA-XYZ* is based upon *XYZ* routing, it is also considered as a minimal routing where each flit from any source and destination pair traverses the minimal number of hops.

5.5.1.5   Switch Allocator Design

The *sw-req* and *port req* signals are issued from each *Input-port* module, and the given information about the desired output-port, are transmitted to the *Switch-Allocator* module to perform the arbitration between the different requests. When more than two input flits from different input ports are requesting the same output-port at the same time, the *Switch-Allocator* manages to decide which output-port should be granted to which input-port, and when this grant should be allocated. This process is done in parallel with the routing computation done in *Input-port* to form the second pipeline stage.

As indicated in Fig. 5.14, the switch allocator circuit has two output signals: one is *sw-cntrl* and the second one is *grant-out*. *sw-cntrl* contains all the information needed by the crossbar circuit about the scheduling result, as is explained later. On the other hand, the *grant-out* is sent back to the *Input-port* module and gives the grant to the appropriate input-port to send its data to the crossbar before reaching

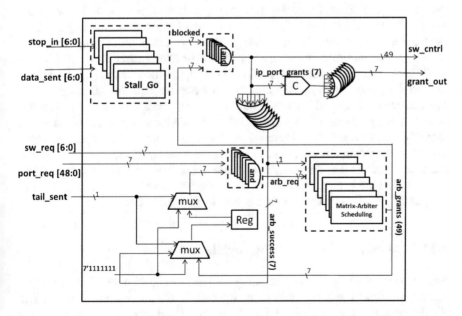

Fig. 5.14 Switch allocator architecture

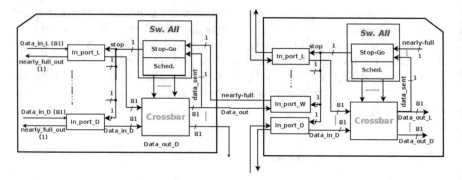

Fig. 5.15 Stall-Go flow control mechanism

its next neighboring node. Figure 5.14 shows that the switch allocator module is composed of two main components: *Stall-Go flow control* and *Matrix-Arbiter Scheduling*.

5.5.1.6   Stall-Go Flow Control Architecture

Like the other flow control schemes, *Stall-Go* module manages the case of the buffer overflow. When the buffer exceeds its limitation on hosting flits (if the number of flits waiting for the process is greater than the depth of the buffer), an efficient flow-control scheme must be considered to prevent buffer overflow and eventually from packet dropping. Thus, allocating available resources to packets as they progress along their route. We chose *Stall-Go* flow control since it proves to be a low-overhead efficient design choice showing remarkable performance compared to the other flow control schemes such us *ACK-NACK* or *Credit based* flow control. The *Stall-Go* module mechanism presented in Fig. 5.15, uses two control signals: *nearly-full* and *data-sent*. *nearly-full* signal is sent to the upstream node indicating that the input buffer is almost full and only one slot is still available to host one last flit. After receiving this signal, the *FIFO* buffers suspend sending flits. The *data-sent* signal is issued when the flit is transmitted. Figure 5.16 represents the *Stall-Go* flow control state machine which aims to generate the *nearly-full* and *data-sent* signals. State *GO* indicates that the buffer is still able to host two or more flits. State *SENT* indicates that the buffer can host only one more flit, and finally, when we move to state *STOP*, it means that the buffer can not store anymore flits.

5.5.1.7   Matrix-Arbiter Scheduling Architecture

The second component is the scheduling module. As shown in Fig. 5.17, the input signals *sw-req* and *port-req* indicate the input-ports that are requesting access, and the output-ports they are requesting respectively. Depending on these requests, the

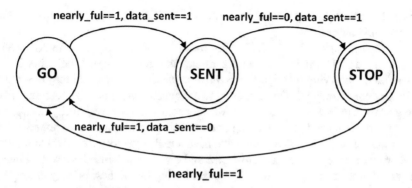

Fig. 5.16 Stall-Go flow control finite state machine

$$
\begin{array}{c|cccccc}
\text{North} & X & 1 & 1 & 1 & 1 & 1 \\
\text{East} & 0 & X & 0 & 0 & 0 & 0 \\
\text{South} & 0 & 1 & X & 1 & 0 & 0 \\
\text{West} & 0 & 1 & 0 & X & 0 & 0 \\
\text{Up} & 0 & 1 & 1 & 1 & X & 0 \\
\text{Down} & 0 & 1 & 1 & 1 & 1 & X
\end{array}
\Longrightarrow
\begin{pmatrix}
X & 0 & 0 & 0 & 0 & 0 \\
1 & X & 0 & 0 & 0 & 0 \\
1 & 1 & X & 1 & 0 & 0 \\
1 & 1 & 0 & X & 0 & 0 \\
1 & 1 & 1 & 1 & X & 0 \\
1 & 1 & 1 & 1 & 1 & X
\end{pmatrix}
\Longrightarrow
\begin{pmatrix}
X & 0 & 0 & 0 & 0 & 1 \\
1 & X & 0 & 0 & 0 & 1 \\
1 & 1 & X & 1 & 0 & 1 \\
1 & 1 & 0 & X & 0 & 1 \\
1 & 1 & 1 & 1 & X & 1 \\
0 & 0 & 0 & 0 & 0 & X
\end{pmatrix}
$$

(a)   (b)   (c)

Fig. 5.17 Scheduling-Matrix priority assignment

arbiter allocates the convenient output-port to the requesting input-port Since 3D-ONoC transmits only one flit in every clock cycle, then when two input-ports or more are competing for the same output-port, the presence of a scheduling scheme is required to prevent any possible conflict. The switch allocator in the design employs a least recently served priority scheme via the packet transmit layer. Thus, it can treat each communication as a partially fixed transmission latency [33, 37]. Matrix arbiter is used for a least recently served priority scheme.

To adopt Matrix arbiter scheduling for 3D-ONoC, we implemented a 6 × 6 scheduling matrix. The scheduling module accepts all the requests from the different connected input-ports and their requested output-ports. Then it assigns priority for each request. To give the grant to the convenient input-port, the scheduling module verifies the scheduling matrix, compares the priorities of the input-ports competing for the same output-port, and initiates the one possessing the highest priority in the matrix. When there are no requests, the priority is unchanged. Based on these assumptions, we are sure that every input-port will be served and get the grant to use the output-port in a fair manner. Figure 5.17 illustrates a simple example of how the scheduling mechanism works. Each row of the matrix represents the competing input requests and their priorities. The scheduling module starts by examining the priority of each input port's request. After the highest priority input is served, the arbiter updates the scheduling matrix by making the input-port which got the last grant, the lowest priority for the next round of arbitration. this is done by inverting its row and column.

The matrix shown in Fig. 5.17a illustrates the initial scheduling-matrix where *North*, *Up* and *Down* input-ports are asking the grant to eject their flits to the *Local* port. Observing this figure, the *North* request (highlighted in red) has higher priorities compared with the remaining two requests. As a result, the Arbiter gives the grant to the *North* request. Then *North* becomes the lowest priority (as a green line underlines it) and the remaining two requests priorities are incremented. In the next round (Fig. 5.17b), *Down* seems to have a higher priority than the *Up* request. The arbiter then gives the grant to *Down* and make its priority the lowest. Finally, as it is shown in Fig. 5.17c, the *Up* request having the highest priority among the others is given the grant to eject its data to the requested output port.

5.5.1.8 Crossbar Design

The switch allocator sends the issued *control* signal to the crossbar circuit to complete the third and final Crossbar Traversal pipeline stage (CT), where information about the selected input port and the *Next-Port* are embedded, and then stored in the *sw-cntrl-reg* register as it is shown in Fig. 5.18. After that, the crossbar fetches this information, receives the data from the FIFO buffer of the selected input port. Then, it allocates the appropriate channel for transmission to the decoded *Next-Port*.

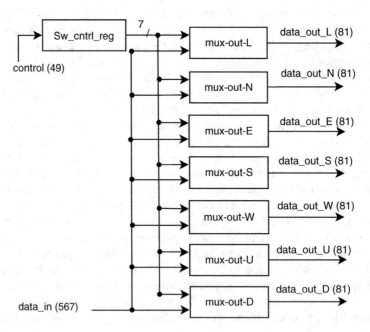

Fig. 5.18 Crossbar circuit

Finally, the crossbar sends the flit to its destination, as illustrated in Fig. 5.18. When all the flits are transmitted, the *tail* bit informs the switch allocator via a *tail-sent* signal that the packet transmission is completed and can free the used channel so it can be exploited by another packet.

5.5.1.9 Network Interface Architecture

To enable real applications to be run on the 3D-ONoC system, a Network Interface (NI) was added to every router as a medium interface between the different PEs (cores, memory, I/O, etc.). JPEG encoder application [78] was used for evaluating the system performance. For this, both *Transmitter* and *Receiver* NI in every switch are designed. The packet size is set to 99-bit (3-bit flits). Each flit contains 17 bits defining the routing information (*xdst*, *ydst*, *zdst*, *Next-Port* and *tail*) and the remaining 16 bits are dedicated for the payload.

Figure 5.19 shows the architecture of the *Transmitter-NI* and Fig. 5.20 shows the architecture of the *receiver-NI*. The NI receives 32 bits data from the JPEG module which will be divided into two portions to represent the payload of the two first flits of the packet. The payload of the third flit contains the 10 bits control signal from the JPEG module and the remaining six bits are unused. As shown

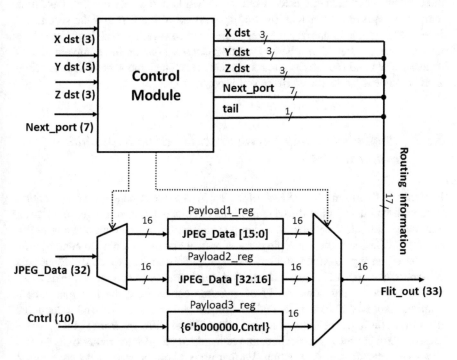

Fig. 5.19 Network interface architecture: transmitter side

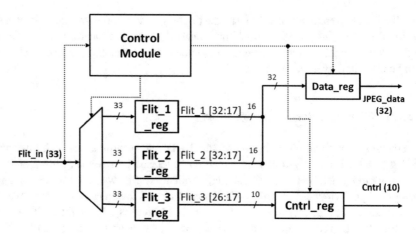

Fig. 5.20 Network interface architecture: receiver side

in Fig. 5.19, a *Control Module* manages flit generation. It adds the convenient destination addresses and *Next-Port* direction to each flit and marks the end of the packet by adding the (*tail* bit to the third final flit. The generated flits are then injected into the network. On the other side, the *Receiver-NI* receives the incoming three flits of each packet ejected from the network, and stores the flits into three temporary registers. After that, the 16 bits payload of the first and the second flit is fetched from the temporary registers, reassembled together, and finally stored in the *Data-reg* register. Controlled by another *Control Module*, the complete 32 bits resulting Data and the 10 bits control signals are fetched and sent to their attached JPEG module after the the entire packet is received.

5.5.2   Photonic On-chip Interconnect for High-Bandwidth Multicore SoCs

Photonic Network-on-Chip (PNoC) [8, 12–15] is a novel concept which enables ultra-high communication throughput in the terabits per second range, low power, and low communication latency. When powered with a wavelength division multiplexing (WDM) scheme, multiple parallel optical streams of data are concurrently transferred through a single on-chip waveguide. This contrasts with the Electronic Networks-on-Chip (ENoC), which requires a unique metal wire per bitstream.

The key to saving power in PNoC systems comes from the fact that once a photonic path is established, the optical data are transmitted end-to-end without the need for buffering, repeating, or regenerating. This is different from ENoCs, where messages are buffered, regenerated, and transmitted on the inter-router links several times en route to their destination. Furthermore, photonic routers do not need to switch to every bit of the transmitted data like electronic routers; optical routers switch on and off once per message. Thus, their energy dissipation does not depend on the bit rate. This feature allows ultra-high bandwidth transmission while avoiding the power cost that is found in traditional ENoCs.

In a hybrid PNoC systems, the source node first issues a path configuration packet, which includes destination address information and other additional control information, via a copper-based electrical link. Next, the configuration packet is routed via an Electric Control Network (ECN), reserving the photonic switches and channels along the path for the photonic message. When the photonic path reservation is completed, the source node returns an Acknowledgment (ACK) signal. When the ACK signal is received and processed by the source node, the optical data transmission starts. At the end of the transmission, all reserved photonic resources for the above data transmission are released.

The circuit-switched nature of such hybrid PNoCs directly affects the performance and power efficiency of on-chip communications. As observed in a previously conducted study, the energy overhead of a hybrid PNoC system is mainly due to the electronic control modules, which consume the majority of the total power budget of a hybrid PNoC system. Moreover, the latency required for photonic path configuration is much longer than the photonic data transfer itself.

While a single-layer configuration can provide low-loss waveguides and high-performance photonic devices, it suffers from limited integration density due to waveguide crossing and limited real estate. A way to go beyond this limitation is to monolithically stack multiple photonic layers above Si as multilayered electrical interconnections realized in modern electronic circuits [10, 91].

Fault tolerance is crucial when considering mission-critical applications where the system must correctly function even when something goes wrong. One such application is space travel, where repair or replacement is not a possible option, and billions of dollars would be wasted.

5.5.2.1 Photonic Communication Building Blocks

The main components of a PNoC include a laser source, which generates phase-coherent and equally spaced wavelengths, waveguides, which is used as a transmission medium, and modulators and photodetectors, which convert digital electrical data to and from photonic signals [48, 51]. It is expected that the laser source could produce up to 64 wavelengths per waveguide for a Dense Wavelength Division Multiplexing (DWDM) network.

Laser Source Since there is no available high-speed, electrically driven, or on-chip monolithic laser light, the PHENIC system features an off-chip laser source, such as VCSEL (Vertically Cavity Surface Emitting Laser). The off-chip laser source provides light to the modulator(s), which transduces electrical information into a modulated optical signal. Then, when the lights enters the chip, optical splitters and waveguides route it to the different modulators used for data transmission.

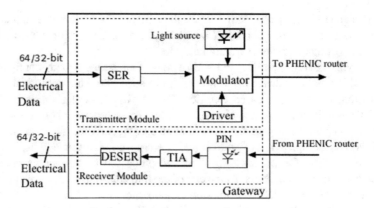

Fig. 5.21 Gateway organization

Modulators Before optical messages are transmitted, the electrical messages from each IP core should be converted to optical form. PHENIC implements at each node a *Gateway:* (Fig. 5.21) serving as a photonic network interface based on silicon optical modulators and SiGe photodetectors. To reduce conversion time, modulators should be small (i.e., the circular-shaped 10 μm ring-modulator [18]) and fast. The performance of a typical modulator is dependent on the on-to-off light intensity ratio [70], which depends on the electrical input signal strength. Therefore, a higher extinction ratio is better and required for fast and accurate signal detection. For example, works in [18, 70] reported that an extinction ratio greater than 10 dB is acceptable and enough to enable proper signal detection without causing communication errors.

Waveguide The waveguides provide the physical interconnection between all sources and destinations, and enable connectivity between all photonic devices in PHENIC systems. The transmitter demultiplexes the light into appropriate wavelength channels and then modulates each channel with a digital data stream generated by the electronic component to be interconnected. Finally, photonic signals are routed to various PEs via routers and waveguides.

We have to note here that the *refractive index* [70] of the waveguide material has a significant impact on the bandwidth, latency, and area of an optical interconnect. A waveguide typically has a width of 0.3 μm [22]. Once the photonic signals are received by the destination node (receiver), the signals must be converted back to electrical form. Also, since PHENIC simultaneously transmits different wavelengths per bidirectional waveguides, a selective wave filter for each received wavelength is needed at the destination node.

Microring Resonator The main element of a silicon photonic NoC system is the microring resonator (MR). MRs can effectively guide an optical signal by carefully choosing their dimensions and positions along the path. Optical signals couple into ring resonators at specific regularly spaced wavelengths in the optical spectrum, called resonant modes [25].

5.5.2.2   Design Challenges

The photonic domain is immune to transient faults caused by radiation [41] but is still susceptible to process variation (PV) and thermal variations (TV) as well as aging. The aging typically occurs faster in active components as well as elements that have high TV [39]. In the optical domain, the faults can occur in MRs, waveguides, routers, etc. Active components, such as MRs, have higher failure rates than passive components, e.g., waveguides [39]. A single MR failure can cause messages to be misdelivered or lost, resulting in bandwidth loss or even complete failure of the whole system. Together, fabrication-induced PV and TV effects present enormous performance and reliability concerns. TV causes a microring to respond to a different wavelength than intended. This can take the form of a passband shift in the MRs. When an MR heats up, it expands, changing its radius and therefore shifting the wavelengths which it uses to the right [21]. As reported in [68], a change of as little as 1 °C can shift the resonance wavelength of a microring by as much as 0.1 nm. This is not permanent and will return when the temperature returns to normal. Therefore, system temperature must be kept at a reasonable value to resonate correctly for the MRs. This is challenging, especially in an extensive complex computing system, which uses thousands of these components. The trimming technique [16] is generally used to dynamically modify the resonance frequency of a microring to overcome both thermal drift and fabrication inaccuracy. This technique can be accomplished by dynamically increasing the current in the $n+$ region or by heating the ring [16, 27, 76].

PV is the variations of critical physical dimensions, e.g., the thickness of the wafer, width of waveguides also affect the resonant wavelengths of MRs. This means that not all fabricated MRs can be used due to PV. As a result, network nodes that do not have all working MRs would lose some or all of wavelengths/bandwidth in communication [85]. To solve this problem, Xu et al. [87] proposed a method of flexible wavelength assignment. Because the networks are already built with excess detectors or Modulators for each message, the node with the excess components can compensate and rematch the components affected by PV.

Over time, all silicon-based ICs wear down. We refer to this phenomenon as *aging*. Some of the aging effects only apply to the active components because of their electrical subcomponents [82], such as the MRs, while other aging affects all parts, even the waveguides.

Recent PNoCs researches (i.e., network topology, router micro-architecture design, and performance and power optimization and analysis) have resulted in several architectures capable of transmitting at a high data bandwidth, and low energy dissipation [8, 12–15]. In [10], we proposed an energy-efficient and high-throughput hybrid silicon-photonic network-on-chip based on a smart contention-aware path-configuration algorithm and an energy-efficient non-blocking optical switch to further exploit the low energy proprieties of the PNoC systems. However, little attention has been given to the aspect of fault-tolerance and reliability along the photonic interconnects.

This section presents a fault-tolerant PNoC architecture. The system is based on a fault-tolerant path-configuration and routing algorithm, a microring fault-resilient photonic router, and uses minimal redundancy to assure the accuracy of the packet transmission even after faulty MRs are detected.

5.5.2.3   Fault Models

It is worth noting that light is not sensitive to radiation or electromagnetic fields, the signals which control optical network can be sensitive to it. The following is a list of actual possible causes that can contribute to the failure of an optical device.

- **PNoC Signal Strength**: Typical NoCs are defined by their power consumption, delay, and throughput. PNoCs also have to consider the Signal-to-Noise Ratio at the receiving end. Because they do not buffer and retransmit, the signal gets weaker based on how many hops it jumps. This does not significantly affect the power the network consumes, but it can lead to a higher sensitivity to noise.
- **Electrostatic Discharge**: While the waveguides are not electrically conductive, the switches and photo-detectors are. This means that they are sensitive to high currents. One thing which can ruin an IC is electrostatic discharge (ESD). This is when a current enters through the I/O pins of the control circuit or can be caused by a strong magnetic field. This all results in the aforementioned extreme current, and this current causes severe damage to the silicon in the components. Possible points of damage are the dielectric, the PN junctions, and any wiring connecting to the controllers. Because of the scaling, the causing phenomena have become harder to control [89]. This can be prevented by providing proper packaging to the IC providing ESD protection at the pins.
- **Noise**: This is one of the unique things that we categorize as a cause for a fault. The reason is that the noise can be caused simply by poorly matched wavelengths. It can also be caused by creating a very-long path or a path that crosses too many intersections. These paths tend to be caused by rerouting or non-minimalistic routing, but other factors can contribute and cause more noise. The most common factors are listed in the following subsections.
- **Aging**: Over time, all silicon-based ICs wear down. Some of the aging effects only apply to the active components because of their electrical subcomponents, while other aging affects the optical properties of the components.
- **Electromigration**: This mainly affects the wires which control the ring resonators. It does not affect the waveguides in any way. It initially causes a delay in the wire and can eventually lead to an open or a short to a nearby wire. It achieves this by thinning out the thinnest portion of the wire due to higher current density at the bottleneck [42].

- **Laser Degradation**: After the lasers have been on for several hundred hours, they start to show signs of degradation. This shows in the form of either missing wavelengths, which can cause a channel fault or general weakening of the original laser signal. However, it does not become an actual problem in each of these cases until the signal falls to a level where the worst-case scenario's Signal-to-Noise ratio is too weak to receive an understandable signal [55].
- **Photodetector Degradation**: Various studies have been done for different types of photodetectors showing that they degrade over time, particularly from being exposed to thermal conditions or UV light. It is reasonable to assume that no matter what material photodetectors are made out of, they all seem to be vulnerable to degradation due to thermal variation, which is present in all networks [39, 82].
- **Aging effect**: Some examples on dealing with aging effects are Agarwal [4], Keane [42], and Kim [43]. These are mainly focused on the electrical side. Many parameters such as the wavelengths and laser strength can possibly be modified throughout the life of a chip to counteract the aging effects similarly to what Mintarno does for Electrical networks [59].
- **Process Variability**: This can affect both the active and inactive components of the optical network. The variability accounts for material impurities, doping concentrations, and size and geometries of structures [75]. For example, one single dimple in a particular point in the coupling region of a ring resonator can significantly affect the coupling properties and thus cause problems for the switch, or maybe just the channel. A poor geometry can also cause a specific component to be more sensitive to aging or ESD. If a variation gets bad enough, an entire link can be rendered useless. This would be considered an early permanent fault and should be detected before a device is released. The impurities in a waveguide can cause such a block or cause a change in the reflectivity of the material, which causes a higher amount of insertion loss, resulting in a lower signal-to-noise ratio. Other similar chains-of-events can occur from lousy doping of the photodetectors. Minimizing this process variability can significantly increase the system's reliability, even without implementing fancier and area or energy-heavy redundancies. The unfortunate truth is that with recent advances in scaling, the variability continues to increase [49, 79].
- **Temperature Variation**: For electrical components, temperature variation can cause changes in properties such as resistivity and generate more power consumption or delay, but in the optical domain, it is quite different. Ring resonators are tuned by heating the ring, causing them to expand, which changes their passband wavelength. If the chip heats up to a point beyond the tuning, then specific channels disappear as a whole. The increase in temperature also causes the photodetectors to degrade, as mentioned in the previous section. These temperature variations also tend to speed up other forms of aging as well.

5.5.2.4   Fault-Tolerant Photonic Network-on-Chip

The Fault-tolerant Photonic Network-on-Chip (FT-PHENIC) system shown in Fig. 5.22, is a mesh-based topology and uses minimal redundancy to assure accuracy of packet transmission even after faulty MRs are detected [57]. The system uses Stall-Go mechanism for flow control and a Matrix-arbiter as a scheduling technique [5, 7, 9]. FT-PHENIC is also based on a microring fault-resilient photonic router (MRPR) and an adaptive path-configuration and routing algorithm [58, 67].

Microring Fault-Resilient Photonic Router

The Microring Fault-resilient Photonic Router (MRPR) consists of a non-blocking fault-tolerant photonic switch and a lightweight control router Redundant MRs are carefully placed at particular locations on the switch to ensure fault tolerance even if one of the MRs on the backup path has a fault. The backup route for the NEWS (North-East-West-South) directions is to use the waveguide connected to the core ports as a master backup; therefore, the redundant MRs are all chosen at locations which connect the NSEW ports to the core.

For most faults, the switch's design allows for an alternate, slightly less power-efficient route. The backup path is less power-efficient because the packets travel across more waveguide distances, go through more active MRs, and cross more waveguides. However, the switch still maintains all of its functionality. Because

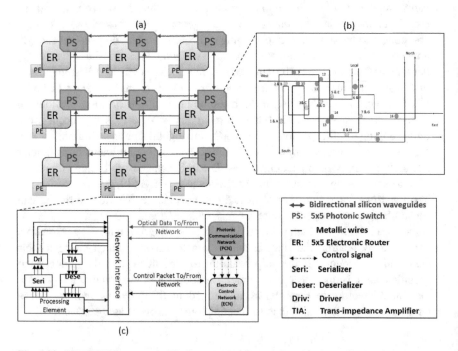

Fig. 5.22 FT-PHENIC system architecture. (**a**) 3 × 3 mesh-based system, (**b**) 5 × 5 non-blocking photonic switch, (**c**) Unified tile including PE, NI and control modules

backup routes are only intended for use in the switches where faults have occurred, the extra loss will have minimal effect on the signal strength of the message across the whole network.

The MRPR was designed to require no MRs from East-West and North-South traffic. Since this kind of traffic accounts for a majority of the traffic of the PCN [58], such design will save on power and continue to function in the case of any MR fails. Assuming that a single location of redundant MRs does not fail altogether, the switch can maintain all functionality at lower speeds.

Figure 5.23 shows a reconfiguration example of how MR 9 can be backed up by MRs 5, 15, and 1. Additionally, the MRs which connect parallel waveguides are replaced with racetracks [60]. This allows for a wider pass-band of light frequencies, making them less sensitive to physical faults, such as reduced sensitivity to thermally-caused passband shifting. Racetracks also have a more considerable mean time between failures (MTBF) [60].

The original MRPR switch is a five-port non-blocking switch, meaning that it allows for routing from any available port to any other available port. Once a fault is detected, the switch recovers, but there is a chance that it may turn into a blocking switch; however, it should maintain all functionality as long as none of the redundant MRs fails. Because the redundant MRs lie dormant, they do not require much power other than the boost in signal strength needed to compensate for the signal loss caused by passing by an inactive MR, which is minimal. However, as all rerouting in the switch occurs on the core waveguide, traffic certainly increases on this single waveguide as too many faults occur, which is why it should be treated as a node failure after a threshold of failed MRs is reached.

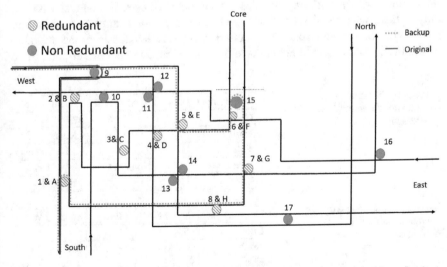

Fig. 5.23 Example of how a non redundant MR's functionality can be mimicked by redundant ones

In addition to tolerating faults, MRPR can handle the *ACK* signals and the resulting regeneration process of the *Tear-down* signal at each hop. To accomplish this goal, a hybrid switching policy is used: *Spacial-switching* for the data signals by manipulating the state of the broadband switching elements and a *Wavelength-selective switching* for the *Tear-down* signals by using detectors and modulators. Moreover, since the *Tear-down* signals should be checked and regenerated at each hop, it is crucial that their manipulation be automatic and not interfere with data signals nor cause a blockage inside the switch. When the *Tear-down* is generated at the source NI (Network Interface), it is first sent to the control router. Then, the *Photonic Switch Controller* releases the corresponding MRs and generates another *Tear-down* which is sent to the output-port modulator in the PCN where it continues its path on a hop-by-hop basis until it reaches its destination. Finally, at the destination node, the *Tear-down* is detected in the input-port and sent to the *Photonic Switch Controller* in the corresponding electronic router. In this fashion, we can omit the overhead of an additional gateway which becomes significant when we increase the number of cores. Table 5.3 shows the MRs configuration for data transmission, where 16 MRs are used in a non-blocking fashion. Table 5.4 shows the backup paths for each transmission.

We use the first six wavelengths in the optical spectrum starting from 1550 nm, with a wavelength spacing equal to 0.8 nm, to maintain a low cross-talk as reported in [73]. For the acknowledgment signals, we use the first five wavelengths in the optical spectrum starting from 1550 nm: four wavelengths for the *Tear-down* signal where each one is dedicated for each port except the local one. In addition, a single wavelength is used for the *ACK*. The remaining available wavelengths are used for data transmission. The five wavelengths used to control the *ACK* and *Tear-down* signals are notably constant regardless of the network size, in contrast with the fully optical, where the number of wavelengths used for control and arbitration grows with the network size. Thus, cutting these wavelengths from the available spectrum for control would not degrade the system bandwidth. Furthermore, these

Table 5.3 Microring configuration for normal data transmission

Output/Input	Core	North	East	South	West
Core	–	4	6	3	5
North	7	–	16	None	14
East	8	17	–	13	None
South	1	None	12	–	9
West	2	11	None	10	–

Table 5.4 Microring backup configuration for data transmission

Output/Input	Core	North	East	South	West
Core	15	D	F	C	E
North	G	–	6,15,7	None	5,15,7
East	H	4,15,8	–	3,15,8	None
South	A	None	6,15,1	–	5,15,1
West	B	4,15,2	None	3,15,2	–

Table 5.5 Wavelength assignment for acknowledgment signal (Mod: Modulator, and Det: Photo-detector)

	Core	North	East	South	West
Input	Mod_{λ_0}	Det_{λ_3}	Det_{λ_2}	Det_{λ_1}	Det_{λ_4}
Output	Det_{λ_0}	Mod_{λ_1}	Mod_{λ_4}	Mod_{λ_3}	Mod_{λ_2}

Table 5.6 Various switches and their estimated losses. AL: Average Loss, WL: Worst Loss

Router	Cros.	MRs	Termi.	AL (dB)	WL (dB)	WL(faulty) (dB)
Crossbar	25	25	10	1.12	1.60	∞
Crux	9	12	2	0.657	1.11	∞
PHENIC	27	18	0	1.315	1.615	∞
FT-PHENIC	19	16+9	0	0.965	1.115	2.215

five wavelengths will be negligible, especially when DWDM is used, providing up to 128 wavelengths per waveguide [23]. The wavelength assignment for each port is shown in Table 5.5.

Should the *Tear-down* signals enter the switch, they need to be redirected to the corresponding electronic router. Since these signals come from different ports and are modulated with different wavelengths, detectors capable of switching all of the four wavelengths are placed in front of the input ports to intercept the signals. The converted optical signal will be redirected to the electronic router to be processed. Then, according to the included information, the corresponding MRs will be released. For the *ACK*, when the PSCP reaches the destination, 1-bit optical signal is modulated from the output port (i.e., opposite direction), to the source. With this intelligent hybrid switching mechanism, we take advantage of the low-power consumption of the optical link by using optical pulses modulated with an adequate wavelength instead of propagating the acknowledgment signals in the ECN. Second, we take advantage of the WDM proprieties by separating the acknowledgment packets and the data signals and letting them coexist in the same medium without interfering with each other. This contrasts with the electronic domain where these acknowledgment packets travel for several hops, consequently blocking (preventing) the waiting cores from sending their PSCP packets. Finally, we can tolerate faults due to the arrangement of the MRs, and allow for redundancy at critical locations.

As a direct comparison, we performed a study on the routers and the loss they would each have on average and in their worst case. The results can be seen in Table 5.6. As expected, the Crux [90] performs the best, as its only design goal was to minimize loss and noise, sacrificing a lot of functionality. Values for the calculation were taken from several authors and can be seen in Table 5.7.

Table 5.7 Insertion loss
parameters for 22 nm process

Parameter	Value
Through ring loss	0.5 dB [90]
Pass by ring loss	0.005 dB [25]
Bending loss	0.005 dB [25]
Crossing loss	0.12 dB [90]
Terminator	0.01 dB [25]

Light-Weight Electronic Control Router

The control router is based upon OASIS-NoC router [1, 3, 7, 9]. The arbiter receives the detected *Tear-down* from an above switch. According to the information encoded in this signal, the corresponding MRs are released, and a new *Tear-down* is generated for the next hop until it reaches its final destination and all MRs involved in this communication are released.The connection between the NI and the local port, enables a configuration packet (CP) to be sent from the NI to the local port. The CP could be a setup packet, or a path blocked packet. The NI is also connected to the data switch (i.e., PCN). When the source node receives the ACK, the payload is processed by a serializer bank (if needed), a high-speed driver, and a modulator to convert the electrical signal to an optical one. The optical data leaves the data switch at the source node and goes through a detection step, a high-speed Trans-Impedance-Amplification step, and a deserialization step. At the end, the NI's receiver receives the payload data with its original clock speed.

Fault-Tolerant Path-Configuration

The key feature of the Fault-tolerant Photonic Path-configuration algorithm (FTPP) is that it can handle faulty MRs within the photonic switches. When a fault occurs, the algorithm checks for the secondary MRs on the list and checks their status. Thus, the backup MR table can be straightforward in the case of a redundant MR failing, where its redundancy replaces it, or it can be slightly more complicated, as seen in Fig. 5.23.

The FTPP algorithm must meet specific requirements to work with the FT-PHENIC system. It should also remove the dependency between the ECN and PCN, which causes a significant latency overhead in conventional hybrid-PNoC systems. In addition, the latency caused by the path blocking, which requires several cycles for the path dropping and the new path setup packet generation, is considerably decreased. Another key feature of the configuration algorithm is the efficiency of the ECN resource utilization. For example, by moving the acknowledgment signals to the upper layer, we can reduce the buffer depth to only two slots since half of the network traffic is eliminated. This reduction is a critical factor in designing a lightweight router, highly optimized for latency and energy.

Before optical data transmission, the source node issues a *Path-setup-Control-Packet* (PSCP), which is routed in the ECN and includes information about the destination and source addresses. In addition to the source and destination addresses, other information is included. For example, one-bit is used for the Packet-type field. This field can be "00" for a PSCP and "01" when this configuration packet is Path-blocked. Other information to ensure Quality-of-Service and fault-tolerance, such as Message-ID, Fault-status, Error-Detection-Code, can also be included.

For each electrical router, the output-port is calculated according to Dimension-Order routing [3]. Every time the PSCP progresses to the next router, the optical waveguides between the previous and current routers are reserved. Depending on the output port of the electrical router, the corresponding photonic router is configured by switching ON/OFF one or more MRs using the MRs configuration table shown in Table 5.3.

5.5.3   Network Interface

In the last two sections, we have introduced the multi-cast and fault-tolerant Network-on-Chip for the neuromorphic system. Finally, a NI (network interface) is needed to allow the neurons to communicate in different nodes.

The inter-neural interconnect consists of multiple routers (R) to handle the communication between the neuron clusters. The inter-neural interconnect supports two types of flit. The first type is the spike between neurons in AER format. The AER format flit is converted to the address of the weight SRAM to feed to the SRAM. The second type of flit is memory access. To read and write the memory cells and registers in the neuron cluster, a flit provides the instruction and the required argument (address). Here, the memory access flits are issued by a master (or external host) processor in the system. We support two types of read/write commands: single and burst. The individual read/write only provides access to one element per request, while an argument of length must follow the burst ones. The NI converts the requested address to the local address at each weight memory or LIF array. Figure 5.24 shows the block diagram of the NI. The input spikes are categorized into either input spikes or memory accesses. With the memory accesses, the NI provides an interface to read and write the data in all registers and memory

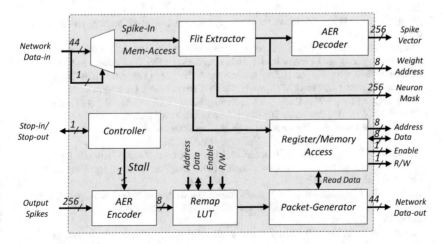

Fig. 5.24 Network interface architecture

blocks of the node. The read instruction makes the NI return the value of the requested address to the master processor. With the input spike from the network, the NI decode phase gets the weight SRAM address and feeds it to the weight memory. For multi-layer SNNs or sparsity connections, the *Flit Extractor* provides the read enable (RE) signal for different layers or different links, which are used in the weight memory. As a result, a node can have multiple AERs at the same address but for other neurons. The LIF array's output spike is fed into the AER decoder, which extracts the address of bit one (firing neuron). This address is then serially sent to the remap Look-Up-Table (LUT) to obtain the AER value in the receiving nodes.

5.6 Chapter Summary

This chapter presented the architecture and circuits used for communication in neuromorphic systems. In particular, the Network-on-Chip fabric is introduced for receiving and transmitting spikes following the Address Event Representation protocol and the memory accessing method. The chapter also covered the interconnect method for inter-neurons communication and the interconnect design principle to help understand the overall concept of on-chip and off-chip communication. Moreover, the chapter introduced advanced on-chip interconnect technologies, including si-photonic three-dimensional interconnects and fault-tolerant routing algorithms.

References

1. Abdallah AB (2013) Multicore systems-on-chip: practical hardware/software design, 2nd edn. Atlantis
2. Abdallah AB (2017) Advanced multicore systems-on-chip: architecture, on-chip network, design. Springer
3. Abdallah AB, Sowa M (2006) Basic network-on-chip interconnection for future gigascale MCSoCs applications: Communication and computation orthogonalization. In Proceedings of Tunisia-Japan symposium on society science and technology (TJASSST), Dec 2006
4. Agarwal M, Paul BC, Zhang M, Mitra S (2007) Circuit failure prediction and its application to transistor aging. In 25th IEEE VLSI test symposium (VTS'07). IEEE, pp 277–286
5. Ahmed AB (2015) High-throughput architecture and routing algorithms towards the design of reliable mesh-based many-core network-on-chip systems. PhD thesis, Graduate School of Computer Science and Engineering, University of Aizu, March 2015
6. Ahmed AB (2016) High-performance scalable photonics on-chip network for many-core systems-on-chip. PhD thesis, GraduteSchool of Computer Science and Engineering, The University of Aizu, March 2016
7. Ahmed AB, Abdallah AB (2013) Architecture and design of high-throughput, low-latency, and fault-tolerant routing algorithm for 3D-network-on-chip (3D-NoC). J Supercomput 66(3):1507–1532
8. Ahmed AB, Abdallah AB (2013) Phenic: silicon photonic 3d-network-on-chip architecture for high-performance heterogeneous many-core system-on-chip. In 2013 14th International

conference on sciences and techniques of automatic control and computer engineering (STA), December 2013, pp 1–9

9. Ahmed AB, Abdallah AB (2014) Graceful deadlock-free fault-tolerant routing algorithm for 3d network-on-chip architectures. J Parallel Distrib Comput 74(4):2229–2240

10. Ahmed AB, Abdallah AB (2015) Hybrid silicon-photonic network-on-chip for future generations of high-performance many-core systems. J Supercomput 71:4446

11. Ahmed AB, Abdallah AB, Kuroda K (2010) Architecture and design of efficient 3d network-on-chip (3d NoC) for custom multicore SoC. In: IEEE Proc. of BWCCA-2010, November 2010, pp 67–73

12. Ahmed AB, Meyer M, Okuyama Y, Abdallah AB (2015) Efficient router architecture, design and performance exploration for many-core hybrid photonic network-on-chip (2d-phenic). In: 2015 2nd International conference on information science and control engineering (ICISCE), April 2015, pp 202–206

13. Ahmed AB, Meyer M, Okuyama Y, Abdallah AB (2015) Hybrid photonic NoC based on non-blocking photonic switch and light-weight electronic router. In: 2015 IEEE international conference on systems, man and cybernetics (SMC), October 2015

14. Ahmed AB, Okuyama Y, Abdallah AB (2015) Contention-free routing for hybrid photonic mesh-based network-on-chip systems. In: The 9th IEEE international symposium on embedded multicore/manycore SoCs (MCSoC), September 2015, pp 235–242

15. Ahmed AB, Okuyama Y, Abdallah AB (2015) Non-blocking electro-optic network-on-chip router for high-throughput and low-power many-core systems. In The World Congress on information technology and computer applications 2015, June

16. Ahn JH, Fiorentino M, Beausoleil RG, Binkert N, Davis A, Fattal D, Jouppi NP, McLaren M, Santori CM, Schreiber RS, Spillane SM, Vantrease D, Xu Q (2009) Devices and architectures for photonic chip-scale integration. Appl Phys A 95(4):989–997

17. Akopyan F, Sawada J, Cassidy A, Alvarez-Icaza R, Arthur J, Merolla P, Imam N, Nakamura Y, Datta P, Nam GJ, Taba B, Beakes M, Brezzo B, Kuang JB, Manohar R, Risk WP, Jackson B, Modha DS (2015) Truenorth: Design and tool flow of a 65 mw 1 million neuron programmable neurosynaptic chip. IEEE Trans Comput Aided Des Integr Circuits Syst 34(10):1537–1557

18. Almeida VR et al (2004) All-optical switching on a silicon chip. Opt Lett 29(24):2867–2869

19. Balaji A et al (2019) Mapping spiking neural networks to neuromorphic hardware. IEEE Trans Very Large Scale Integr (VLSI) Syst 28(1):76–86

20. Bhanu PV et al (2019) Fault-tolerant network-on-chip design with flexible spare core placement. J Emerg Technol Comput Syst 15(1):1

21. Bogaerts W, De Heyn P, Van Vaerenbergh T, De Vos K, Kumar Selvaraja S, Claes T, Dumon P, Bienstman P, Van Thourhout D, Baets R (2012) Silicon microring resonators. Laser Photonics Rev 6(1):47–73

22. Briere M et al (2005) Heterogeneous modelling of an optical network-on-chip with systemc,. In: 16th IEEE international workshop on rapid system prototyping (RSP'05), June, pp 10–16

23. Brusberg L, Schröder H, Queisser M, Lang KD (2012) Single-mode glass waveguide platform for dwdm chip-to-chip interconnects. In: 2012 IEEE 62nd electronic components and technology conference (ECTC), May, pp 1532–1539

24. Carloni LP, Pande P, Xie Y (2009) Networks-on-chip in emerging interconnect paradigms: Advantages and challenges. In: Proceedings of the 3rd ACM/IEEE international symposium on networks-on-chip (NOCS09), May, pp 93–102

25. Chan J, Hendry G, Biberman A, Bergman K, Carloni LP (2010) Phoenixsim: A simulator for physical-layer analysis of chip-scale photonic interconnection networks. In: Proceedings of the conference on design, automation and test in Europe. European Design and Automation Association, pp 691–696

26. Chao CH, Jheng KY, Wang HY, Wu JC, Wu AY (2010) Traffic- and thermal-aware run-time thermal management scheme for 3d NoC systems. In Proc. ACM/IEEE Int. Symp. Networks-on-Chip (NoCS), May, pp 223–230, Grenoble

27. Chu ST, Pan W, Sato S, Kaneko T, Little BE, Kokubun Y (1999) Wavelength trimming of a microring resonator filter by means of a uv sensitive polymer overlay. IEEE Photonics Technol Lett 11(6):688–690

28. Dally WJ (1991) Express cubes: Improving the performance of kary-n-cube interconnection networks. IEEE Trans Comput 40(9):1016–1023
29. Das S et al (2004) Technology, performance, and computer aided design of three-dimensional integrated circuits. In: In Proc. international symposium on physical design
30. Davies M, Srinivasa N, Lin TH, Chinya G, Cao Y, Choday SH, Dimou G, Joshi P, Imam N, Jain S, et al (2018) Loihi: A neuromorphic manycore processor with on-chip learning. IEEE Micro 38(1):82–99
31. Dev K (2002) Multi-objective optimization using evolutionary algorithms. Wiley
32. Feero B, Pande PP (2007) Performance evaluation for three-dimensional networks-on-chip. In: Proceedings of IEEE Computer Society annual symposium on VLSI (ISVLSI), May, pp 305–310
33. Z Fu, Ling X (2010) The design and implementation of arbiters for network-on-chips. In: 2nd International conference industrial and information systems, pp 292–295
34. Furber S, Temple S (2007) Neural systems engineering. J R Soc Interface 4(13):193–206
35. Furber SB, Lester DR, Plana LA, Garside JD, Painkras E, Temple S, Brown AD (2013) Overview of the spinnaker system architecture. IEEE Trans Comput 62(12):2454–2467
36. Glass CJ, Ni LM (1992) The turn model for adaptive routing. In: Proc. 19th Ann. Int'l Symp. computer architecture, May, pp 278–287
37. Gold BT (2004) Balancing performance, area, and power in an on-chip network. Master's thesis, Department of Electrical and Computer Engineering, Virginia Tech, August 2004
38. Habibi A, Arjomand M, Sarbazi-Azad H (2011) Multicast-aware mapping algorithm for on-chip networks. In: 19th International euromicro conference on parallel distributed and network-based processing, February 2011, pp 455–462
39. Hu ZS, Hung FY, Chen KJ, Chang SJ, Hsieh WK, Liao TY (2013) Improvement in thermal degradation of zno photodetector by embedding silver oxide nanoparticles. Funct Mater Lett 6(01):1350001
40. Joyner J, Zarkesh-Ha P, Meindl J (2001) A stochastic global net-length distribution for a three-dimensional system-on-chip(3d-soc). In Proc. 14th annual IEEE international ASIC/SOC conference, September
41. Kappeler R (2004) Radiation testing of micro photonic components. Stagiaire project report. Technical report, ESA/ESTEC. September 29 Ref. No.: EWP 2263
42. Keane J, Kim CH (2011) An odometer for cpus: Microprocessors don't normally show wear and tear, but wear they do. IEEE Spectr 48(5):26–31
43. Keane J, Kim TH, Kim CH (2010) An on-chip NBTI sensor for measuring PMOS threshold voltage degradation. IEEE Trans Very Large Scale Integr (VLSI) Syst 18(6):947–956
44. Kelber F, Wu B, Vogginger B, Partzsch J, Liu C, Stolba M, Mayr C (2020) Mapping deep neural networks on spinnaker2. In: Proceedings of the neuro-inspired computational elements workshop, pp 1–3
45. Kim K, Kim HY, Kim TG (2003) Top-down retargetable framework with token-level design for accelerating simulation time of processor architecture. IEICE Trans. fundamentals of electronics, communications and computer sciences, December, Vol. E86-A,(12), pp 3089–3098
46. Kim J, Nicopoulos C, Park D, Narayanan V, Yousif MS, Das CR (2006) A gracefully degrading and energy-efficient modular router architecture for on-chip networks. In: Proc. of the 33rd Int. Sym. on Comp. Arch, pp 138–149
47. Kim J, Balfour J, Dally WJ (2007) Flatterned butterfly topology for on-chip networks. In Proc. of the 40th Int. Sym. on microarchitecture, pp 172–182
48. Koch BR, Fang AW, Cohen O, Bowers JE (2007) Mode-locked silicon evanescent lasers. Optics Express 15(18):11225–11233
49. Kuhn K, Kenyon C, Kornfeld A, Liu M, Maheshwari A, kai Shih W, Sivakumar S, Taylor G, VanDerVoorn P, Zawadzki K (2008) Managing process variation in intel's 45nm cmos technology. Intel Technol J 12:2
50. Kumar R, Zyuban V, Tullsen DM (2005) Interconnections in multicore architectures: Understanding mechanisms, overheads and scaling. In: Proc. of the 32nd Int. Sym. on Comp. Arch, Madison, pp 408–419

51. Kumar A, Peh LS, Kundu P, Jha NK (2007) Express virtual channels: Towards the ideal interconnection fabric. In: Proc. of the 34th Int. Sym. on Comp. Arch, pp 150–161
52. Lahiri K, Raghunathan A, Dey S (2000) Efficient exploration of the SoC communication architecture design space. In: Proc. IEEE/ACM ICCAD'00, pp 424–430
53. Leary G, Chatha KS (2010) Design of NoC for SoC with multiple use cases requiring guaranteed performance. In: 23rd International conference on VLSI design, January, pp 200–205
54. Li F, Nicopoulos C, Richardson T, Xie Y, Narayanan V, Kandemir M (2006) Design and management of 3d chip multiprocessors using network-in-memory. ACM SIGARCH Comput Architect News 34(2):130–141
55. Luryi S, Xu J, Zaslavsky A (2007) Future trends in microelectronics: up the nano creek. Wiley John & Sons
56. Mahowald M (1992) VLSI analogs of neuronal visual processing: a synthesis of form and function. PhD thesis, California Institute of Technology Pasadena
57. Meyer M (2017) Micro-ring fault-resilient photonic on-chip network for reliable high-performance many-core systems-on-chip. PhD thesis, Graduate School of Computer Science and Engineering, The University of Aizu, March 2017
58. Meyer MC, Ahmed AB, Okuyama Y, Abdallah AB (2015) Fttdor: Microring fault-resilient optical router for reliable optical network-on-chip systems. In: 2015 IEEE 9th international symposium on embedded multicore/many-core systems-on-chip (MCSoC), September, pp 227–234
59. Mintarno E, Skaf J, Zheng R, Velamala JB, Cao Y, Boyd S, Dutton RW, Mitra S (2011) Self-tuning for maximized lifetime energy-efficiency in the presence of circuit aging. IEEE Trans Comput Aided Des Integr Circuits Syst 30(5):760–773
60. Mohamed M (2013) Silicon nanophotonics for many-core on-chip networks. PhD thesis, University of Colorado
61. Montana JM, Koibuchi M, Matsutani H, Amano H (2009) Balanced dimension-order routing for k-ary n-cubes. In International conference on parallel processing workshops, pp 499–506
62. Moradi S, Qiao N, Stefanini F, Indiveri G (2017) A scalable multicore architecture with heterogeneous memory structures for dynamic neuromorphic asynchronous processors (dynaps). IEEE Trans Biomed Circuits Syst 12(1):106–122
63. Mori K, Abdallah AB, Kuroda K (2009) Design and evaluation of a complexity effective network-on-chip architecture on FPGA. In: The 19th intelligent system symposium (FAN 2009), September, pp 318–321
64. Mori K, Esch A, Abdallah AB, Kuroda K (2010) Advanced design issue for oasis network-on-chip architecture. In: International conference on BWCCA. IEEE, pp 74–79
65. Morrow P, Kobrinsky M, Ramanathan S, Park CM, Harmes M, Ramachandrarao V, Park H, Kloster G, List S, Kim S (2004) Wafer-level 3d interconnects via cu bonding. In: Proc. the 21st advanced metallization conference, October
66. Mullins R, West A, Moore S (2004) Low-latency virtual-channel routers for on-chip networks. In: Proc. of the 31st Int. Sym. on Comp. Arch, pp 188–197
67. Nikdast M, Nicolescu G, Beux SL, Xu J (2017) Photonic interconnects for computing systems. River Publishers Series. ISBN-13: 9788793519800
68. Nitta CJ, Farrens MK, Akella V (2011) Resilient microring resonator based photonic networks. In: Proceedings of the 44th annual IEEE/ACM international symposium on microarchitecture, New York, NY, ACM MICRO-44, pp 95–104
69. Orgas UY, Marculescu R (2006) It's a small world after all: NoC performance optimization via long-range link insertion. IEEE Trans VLSI Syst 14(7):693–706
70. Parsricha S, Dutt N (2008) Trends in emerging on-chip interconnect technologies,. IPSJ Trans Syst LSI Des Methodol 1:2–17
71. Pavlidis VF, Friedman EG (2007) 3-d topologies for networks-on-chip. IEEE Trans VLSI Syst 15:1081–1090
72. Philip G, Christopher B, Ramm P (2008) Handbook of 3D integration: technology and applications of 3D integrated circuits. Wiley-VCH

73. Preston K, Sherwood-Droz N, Levy JS, Lipson M (2011) Performance guidelines for WDM interconnects based on silicon microring resonators. In: 2011 Conference on lasers and electro-optics (CLEO), May, pp 1–2
74. Purves D, Augustine G, Fitzpatrick D, Hall W, LaMantia AS, McNamara J (2008) Neuroscience. Sinauer Associates
75. Radetzki M, Feng C, Zhao X, Jantsch A (2013) Methods for fault tolerance in networks-on-chip. ACM Comput Surv (CSUR) 46(1):8
76. Rafizadeh D, Zhang JP, Hagness SC, Taflove A, Stair KA, Ho ST, Tiberio RC (1997) Temperature tuning of microcavity ring and disk resonators at 1.5-/spl mu/m. In: Conference proceedings. LEOS'97. 10th annual meeting IEEE lasers and electro-optics society 1997 annual meeting, November, pp 162–163
77. Ramanujam RS, Lin B (2008) Near-optimal oblivious routing on three dimensional mesh networks. In: Proc. IEEE Int. Conf. Comp. Design, Lake Tahoe, CA, pp 134–141
78. Rosethal J (2006) Jpeg image compression using an FPGA. Master's thesis, Electrical and Computer Engineering, University of California Santa Barbara, December 2006
79. Saha SK (2010) Modeling process variability in scaled CMOS technology. IEEE Des Test Comput 27(2):8–16
80. Sivilotti MA (1991) Wiring considerations in analog VLSI systems, with application to field-programmable networks. PhD thesis, California Institute of Technology
81. Topol AW, La Tulipe DC, Shi L, Frank DJ, Bernstein K, Steen SE, Kumar A, Singco GU, Young AM, Guarini KW, Ieong M (2006) Three-dimensional integrated circuits. IBM J Res Devel 50(4/5):491–506
82. Tu Z, Zhou Z, Wang X (2014) Reliability considerations of high speed germanium waveguide photodetectors. In: Optical components and materials XI, pp 89820W–89820W
83. Tyagi S (2009) Extended balanced dimension ordered routing algorithm for 3d-networks. In: International conference on parallel processing workshops, pp 499–506
84. Urgese G, Barchi F, Macii E, Acquaviva A (2016) Optimizing network traffic for spiking neural network simulations on densely interconnected many-core neuromorphic platforms. IEEE Trans Emerg Top Comput 6(3):317–329
85. Xiang D, Zhang Y, Shan S, Xu Y (2013) A fault-tolerant routing algorithm design for on-chip optical networks. In: 2013 IEEE 32nd international symposium on reliable distributed systems (SRDS), September, pp 1–9
86. Xin L, Choy CS (2010) Low-latency NoC router with lookahead bypass. In: IEEE Int Symp. on circuits and systems, pp 3981–3984
87. Xu Y, Yang J, Melhem R (2012) Tolerating process variations in nanophotonic on-chip networks. ACM SIGARCH Comput Architect News 40:142–152
88. Yan S, Lin B (2008) Design of application-specific 3d networks-on-chip architectures,. In: Proceedings of international conference of computer design, October, pp 142–149
89. Yang SG, Li L, Xu Y, Zhang YA, Zhang B (2007) A power-aware adaptive routing scheme for network on a chip. In: 7th International conference on ASIC, pp 1301–1304
90. Ye Y, Wu X, Xu J, Zhang W, Nikdast M, Wang X (2012) Holistic comparison of optical routers for chip multiprocessors. In: 2012 International conference on anti-counterfeiting, security and identification (ASID). IEEE, pp 1–5
91. Zhu S, Lo GQ (2015) Vertically-stacked multilayer photonics on bulk silicon toward three-dimensional integration. J Lightwave Technol 34(2):386–392

Chapter 6
Fault-Tolerant Neuromorphic System Design

Abstract Neuromorphic computing systems have shown tremendous progress in many real-world applications (i.e., object recognition, robotics, autonomous vehicles, etc.). To develop such emerging systems, designers use large-scale models on dedicated hardware platforms, such as FPGAs, GPUs, or ASICs. The designers need a long time to collect datasets, train, and design accelerators to keep the trained models private and reliable. However, with the growing complexity of neuromrphic systems, there are severe vulnerabilities in the hardware implementations. An attacker who does not know the details of structures and designs inside these accelerators can effectively reverse engineer the neural networks by leveraging various side-channel information. Moreover, as neuromorphic systems are complex and integrate large number of neurons and synapses, the fault probability is accumulated and can threaten system reliability. This chapter covers the main threats of reliability and discusses several recovery methods.

6.1 Introduction

When manufacturing Integrated Circuits (ICs), there is a specific value of yield rate that is exceptionally critical [20]. Inaccuracy in the fabrication process can lead to several variations from the original design. Consequently, the fabricated devices are not always perfect. If the inaccuracy leads to a mistake in the functionality of the system, a *fault* is considered to happen. For example, consider the threshold comparison circuit that helps dedicate the neuron's firing status with the output line stuck at '1' (high voltage); the neuron is constantly firing regardless of its inputs or synapses' strength. This type of fault causes an error when the neuron's output is used to drive other neurons that indicate the output and change the correct outcome (i.e., a different neuron with a different label keeps firing).

Other aspects that need to be considered are the wear-out or aging processes. Assuming the system is fabricated correctly and can perform as precisely as it is designed, Wear-out or aging can occur, leading to erroneous outputs. For example, the wires connecting two clusters of neurons can generally work at the beginning. However, due to a constant current on the wires, the *electromigration* effect causes a gradual movement of the ions of the conductor and keeps thinning the wires. After a certain number of operating hours, the wires can be disconnected and no longer transmit the spikes between two clusters.

Hardware faults can be classified according to different aspects. They can be classified by the duration into *permanent*, *intermittent*, and *transient*. A *permanent* fault occurs constantly and never return to be functional. A *transient* fault causes a component malfunction some time and can go away after a short period. A *permanent* fault does not go away, but it usually oscillates. If the fault is active, the component malfunctions, and if the fault is inactive, the component works normally.

Another classification for faults is by their behaviors. For instance, a fault on a resistive memory cell can be classified as: *stuck-at-ground*, *stuck-at-supply-voltage*, *stuck-at-high-resistance* and *stuck-at-low-resistance*. When a resistive memory cell is *stuck-at-ground*, the measured voltage at the reading terminal is stuck at the ground voltage (0 V). This behavior does not allow the resistance value can be read and executed correctly.

As the computation of neuromorphic systems are based on spikes, they can be resilient against transient faults as the impact of faulty spikes can be alleviated. On the other hand, *permanent* and *intermittent* faults can cause malfunction in modules of neuromorphic systems, leading to inaccurate results.

6.1.1 Measure of Fault Tolerance

Because the reliability of a system can be critical, it is important to have a proper measurement method. The traditional approach is to measure *reliability* and *availability*. *Reliability*, denoted as $R(t)$, is the probability that the system work normally in the interval $[0, t]$. A closely related measurement is the *Mean Time to Failures* (MTTF). It is the average time between two consecutive failures and can be computed as:

$$MTTF = \int_0^\infty R(t)dt \qquad (6.1)$$

Another measurement for repairable systems is the *Mean Time Between Failures* (MTBF) which is the sum of *Mean Time to Failures* and *Mean Time to Repair* (MTTR).

Availability denoted as $A(t)$ is the average fraction of time over the interval $[0, t]$ that the system is working. The long-term Availability (A) can be computed as:

$$A = \lim_{0 \to \infty} A(t) = \frac{MTTF}{MTBF} = \frac{MTTF}{MTTF + MTTR} \qquad (6.2)$$

The fault rate ($\lambda(t)$) of a system, considered as the inverted value of $R(t)$, can also be measured by the sum of all fault rates of all modules in the system.

$$\lambda_{system}(t) = \sum_{all-modules} \lambda_{module_i}(t) \qquad (6.3)$$

$$R(t) = \frac{1}{\lambda_{system}(t)} = \frac{1}{\sum_{all-modules} \frac{1}{Rmodule_i(t)}} \qquad (6.4)$$

If redundancies are added, the computation of fault-rate can be changed. The standard method is to use the *Markov-state* model to calculate the probability of faulty states, and $\lambda(t)$ is the sum of them [21].

6.1.2 Type of Faults and Behavior

From the reliability perspective, a typical neuromorphic system consists of three major parts: (1) *memory* (or storage unit), (2) *computing unit*, and (3) *communication infrastructure*. Table 6.1 depicts the faults in terms of type, behavior, detection and recovery approach. In this section, we cover the hardware fault-tolerance method only. In *memory*, particularly SRAM and DRAM, the common faults are *transient* faults, caused by alpha particles or cosmic rays [3, 33]. Two transistors hold the state of an SRAM cell, and with DRAM, it is a capacitor. Alpha particles can switch the cell's value that flips the bit value (0 to 1, 1 to 0), which is called a single-event upsets (SEU). Since the causes of this type of error are unpreventable and unpredictable, it can only be dealt with by using an information redundancy method (i.e., repetition or error correction codes). Another common type is *permanent* faults [6] where the value of the cell is stuck at 0 or 1. *Intermittent* faults can also occur in the memory cells and be active at a certain condition such as thermal elevation [29]. There are certain damages that can occur within the memory, and the common recovery approach is to replace the cells/banks/blocks with spare ones. Faults on the arbitration modules (reading/writing sub-module) can also cause malfunctions. This type of fault can lead to inaccurate reading and writing results.

Faults on *computing unit* are likely to corrupt the output of the circuit. As we mentioned earlier, a *permanent* fault such as *stuck-at-one* on the output wire of the threshold comparator leads to a constant firing neuron. To detect and recover from *permanent defects*, having redundancy and performing voting is necessary. The faults can also be *intermittent* and active under a specific condition. Similar to SEU, single-event transients (SET) can occur in the combinational logic that alternates the circuit's output in a short interval. Since neuromorphic systems can be resilient to some noisy inputs, it can also be resilient to a certain amount of transient faults. If the transient errors are too frequent and can affect the overall accuracy, having redundancies and a voting circuit can help solve the problem.

Table 6.1 Taxonomy of faults: types, causes, behaviors, detection and recovery

Major part	Type	Causes	Behavior	Detection method	Recovery
Memory	Transient	Alpha particles/cosmic rays	Flip bit	Replicating and comparing/error detection code	Information redundancy
	Permanent/intermittent	Manufacture imperfection/aging/wear-out	Stuck-at, bridge	Testing algorithm	Spatial redundancy
Computing unit	Transient	Alpha particles/cosmic rays	Inaccurate output	Redundancy-based voting/multiple executions	Self-resilient/redundancy-based voting/multiple executions
	Permanent/intermittent	Manufacture imperfection/aging/wear-out	Stuck-at, bridge	Voting	Spatial redundancy
Communication infrastructure	Transient	Alpha particles/cosmic rays	Corrupted data, mis-routing	Error detection code/multiple executions	Error correction code/network re-routing
	Permanent/intermittent	Manufacture imperfection/aging/wear-out	Corrupted data, blocking connection	Error detection code	Spatial redundancy, fault-tolerant routing

While the faults on *communication infrastructure* is similar to *computing units*, their behavior are different. *Transient* faults can cause data corruption or misrouting of packets. Error detection and error correction codes can be used to detect and correct the corrupted data [8]. For misrouted packets, the *communication infrastructure* need to detect and re-route them [9]. Dropping packets with the help of flow-control protocol can ensure the sender resend them. *Intermittent and permanent* faults on the routing unit can corrupt data or completely block connections. Having spatial redundancies (i.e., extra wires) or fault-tolerant routing (to avoid blocked connections) can be useful.

While *fault-tolerance* methods for computer systems are well mature and has been developed for decades, *neuromorphic computing* has recently become the new computing approach. Therefore, adopting and adapting the existing methods can be helpful. Furthermore, having a dedicated strategy for neuromorphic systems is also necessary. Details on how to tolerate faults in neuromorphic systems will be further discussed in the following sections.

6.1.3   Impact of Faults on Neuromorphic System

To understand the impact of faults on SNN, we randomly inserted several faults and tested with the 10,000 test cases in the MNIST (Modified National Institute of Standards and Technology database) dataset. The SNN model is 784:100 with lateral inhibitory connections adopted from [11] and run on BindsNet [13] simulator. This network follows the winner-take-all principle, where a firing neuron inhibits other neurons. The weights are pre-trained using STDP algorithm as in [11]. Figure 6.1 illustrates the accuracy drop when inserting faults into the weight SRAM. Once we inserted the stuck-at-0 faults into the weight memory, as shown in Fig. 6.1, we notice that the accuracy drops are ineligible for a small number of faulty weights thanks to the natural fault-resilience of SNNs. However, when the number of faults

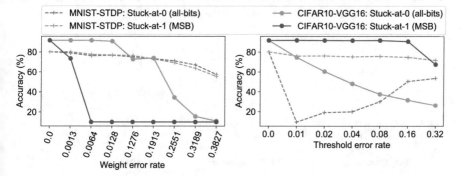

Fig. 6.1 Impact of faults on a neuromorphic system

increases, the accuracy starts dropping significantly. On the other hand, a defect in a computation unit is more critical. For example, a single stuck-at-0 (1% in 100-neurons) on a threshold register can easily make a constant firing neuron, which drops the accuracy significantly to around 10%. This is equal to assigning one label for all testing images of MNIST. Once two or more neurons have stuck-at-0 faults at their threshold registers, two or more neurons start to compete, which increases the overall accuracy. However, the accuracy is still much lower than the non-faulty results.

The impact on faults is also critical in deep neural networks. For example, with the VGG-16 model on CIFAR-10, the accuracy is also significantly dropped while inserting faults.

6.2   Conventional Computing System Fault Tolerance

This section covers the conventional fault-tolerance approach for computing systems. As these fault tolerance approaches can be re-applied for neuromorphic computing systems, this section aims to present an overview of existing techniques. Fault-tolerance techniques that are dedicated to neuromorphic systems are illustrated in the next section.

To summarize the conventional methods, this section covers three major approaches: (1) hardware approach, (2) information redundancy, and (3) software approach.

6.2.1   Hardware Approach

The hardware approach is the most mature one in the field of fault-tolerant computing. Before analyzing the reliability of hardware structure, connecting the module is important as it can affect the overall reliability. In short, there are three types of structure: *parallel*, *serial* and *mixed* as shown in Fig. 6.2.

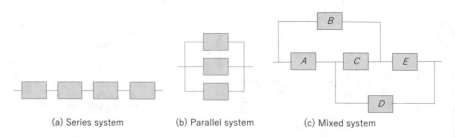

Fig. 6.2 Impact of faults on a neuromorphic system. (**a**) Series system. (**b**) Parallel system. (**c**) Mixed system

In the *serial* system, the modules are connected. If one of the modules fails, the whole system will malfunction. Assuming the modules fail independently, the reliability of the system is the product of the reliability of all modules:

$$\lambda_{system}(t) = \prod_{all-modules} \lambda_{module_i}(t) \tag{6.5}$$

In the *parallel* system, the system only fails when all modules are failed. Consequently, the reliability of the parallel system is:

$$\lambda_{system}(t) = 1 - \prod_{all-modules} (1 - \lambda_{module_i}(t)) \tag{6.6}$$

For the *mixed* system, we can divide it into sub-systems to analyze. However, it might become overwhelmingly complicated. Here, the efficient approach is to build the Markov-state model. The model consists of all possible states of a system that might occur. Between states, there are links with a certain probability. By finding the probability of the faulty condition, the reliability of a system can be obtained.

In summary, the *parallel* system is usually used for tolerating faults. Here, spare modules are added to allow the system to work under one or more faulty modules. The generalized model of this fault-tolerance method is called *M-of-N* systems. Here, the system consists of N modules, and the system requires at least M modules to work properly. For example, one of the most popular fault-tolerant systems is TMR (Triple Modular Redundancy), where a module is tripled, and the system can tolerate at least one failed module. Figure 6.3 depicts the overall structure of a TMR. TMR can be considered as a 2-of-3 system. The reliability function of the M-of-N system is:

$$R_{\text{M-of-N}}(t) = \sum_{i=M}^{N} (Ni) R^i(t)[1 - R(t)]^{N-i} \tag{6.7}$$

Since the voter is connected serially to the three parallel modules, the actual system reliability is

$$R_{system}(t) = R_{\text{M-of-N}}(t) R_{voter}(t) \tag{6.8}$$

$$= R_{voter}(t) \sum_{i=M}^{N} (Ni) R^i(t)[1 - R(t)]^{N-i} \tag{6.9}$$

Fig. 6.3 Triple modular redundancy

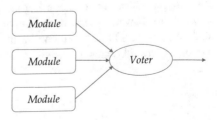

The voter becomes the most critical module in this system as its failure cannot be detected and corrected.

There are also variations of M-of-N with voter, and they include:

- *Sub-system TMR:* instead of having TMR for the whole system, each module is replicated three times. There are also three voters to conduct the voting system. The output of three voters is brought to the next module. This can avoid faulty voters among the system as it can be corrected.
- *Dynamic redundancy:* redundant modules are inactive during operation. Once the voter or fault detection module finds faulty output, the system reconfigures by replacing the failed module with a spare one.

Besides adding spare modules for correction, these spares can also be used for replication for a comparison that will indicate whether there is a fault in them. On processor-based computing, there are other similar approaches for predicting failure situations. To ensure the processor keeps working and does not crash, *watchdog* processor is used with assertions during operation. This approach is also used in other computing approaches as *Runtime detection* that can indicate failures. For multi-core systems, *executing parallel threads* of the same copy of the program can also help show possible errors.

In summary, the hardware fault tolerance approach generally relies on having redundancy to detect and correct failed modules. This approach can be applied to neuromorphic systems. For example, instead of having 256 working neurons in a cluster, the system can add 16 spare neurons to have a 256-of-272 cluster. Once a neuron fails, the spare neurons can be used as replacements. The TMR method can also be applied to computing units in the neuron. As we mentioned that a fault in the threshold comparator could lead to faulty output, having three comparators and a voter can help detect whether a neuron fires or not.

6.2.2 Information Redundancy

While adding redundancies and monitors are common in hardware fault-tolerance, having them in data is not efficient. For instance, using TMR for data triples the amount of storage in the system. To solve this issue, having a proper *information redundancy* is needed. The most common form is *coding* where the data is encoded to a codeword. Then, the codeword will be stored and decoded to detect and correct possible errors. When *encoding*, a d-bit data word is transformed to a c-bit codeword $(c > d)$. This introduces the redundancy (c–d bits) in the information. Note that the possible codewords do not cover all 2^c binary combinations, which leave some combinations invalid. While decoding, the system can encounter a valid or an invalid codeword. A valid codeword can be transformed to obtain the original d-bit data word. An invalid codeword needs to be considered to ensure the possible original data.

Fig. 6.4 A 3-bit codeword space. Green box: invalid codeword. Read box: invalid codeword

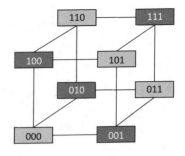

An important metric of the codeword space is the *Hamming* distance which is determined as the number of different bits between two codewords. For example, between "110" and "011" in Fig. 6.4, the Hamming distance are two. This *Hamming* distance helps determine the ability to detect and correct errors. To have the ability to detect k error bits, the *minimum* Hamming distance must be at least $k + 1$. The minimum distance in the previous example is two; therefore, this codeword space can detect up to 1 error bit. For example, if we receive a codeword "010", we can observe that this codeword has the Hamming distance to "110" and "011". Therefore, once the decoder receives "010", it can detect there is an error bit; however, it cannot indicate the original codeword. To detect the original codeword, the minimal Hamming distance must be higher. To correct k bits, the minimal distance must be $2k + 1$. Once we receive an invalid codeword, we can determine which codeword is closest and determine it as the original one. For example, using TMR for one bit can lead to similar codeword space in Fig. 6.4 with only two codewords, "000" and "111" are valid. The minimal Hamming distance is now three. Therefore, it can correct 1 error bit. If the decode receive a codeword "010", it can determine the original codeword as "000" due to the *Hamming* distance to "000" and "111" are 1 and 2, respectively.

6.2.2.1 Parity Code

One of the most basic coding techniques is *parity* code. The parity codeword consist of the d-bit data and one parity bit ($c = d + 1$). The parity bit is the output of the parity check of the d-bit data word. The parity check can be performed using the XOR function and can be designed with several XOR gates in hardware as shown in Fig. 6.5. Therefore, this *coding* technique is simple and fast.

Parity code has two type: *odd* and *even*. The *even* and *odd* parity bit make the codeword to have output parity of '0' and "1", respectively. For example, the 8-bit data is "01010110" as in Eq. 6.10. The *odd* and *even* parity bit is '1' and '0', respectively.

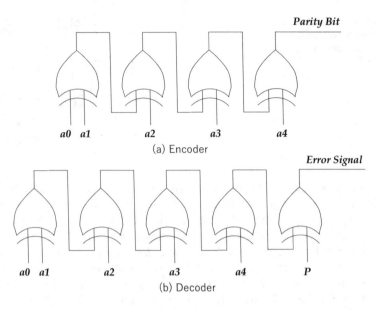

Fig. 6.5 Even parity code: (**a**) encoder; (**b**) decoder

The minimal Hamming distance is two; therefore, *Parity* code can detect one error bit.

$$\text{Data}: 0\ 1\ 0\ 1\ 0\ 1\ 1\ 0$$

$$\text{Even parity codeword}: 0\ 1\ 0\ 1\ 0\ 1\ 1\ 0\ \mathbf{0} \tag{6.10}$$

$$\text{Odd parity codeword}: 0\ 1\ 0\ 1\ 0\ 1\ 1\ 0\ \mathbf{1}$$

6.2.2.2 Hamming Code

Another basic coding technique is *Hamming* code which can correct one bit. There is another variation called SECDED (single error correction, double error detection), which can detect two error bits as its name suggests.

Hamming code has $2^r - r - 1$ bit data word and $2^r - 1$ codeword length ($r \geq 2$). The SECDED code has an extra bit, so the codeword is 2^r bit. An example of Hamming code parity combinations is shown in Table 6.2. Here, Hamming code can be considered as several parity codes of different parts of a data word. The minimal Hamming distance of Hamming code is 3; therefore, it can correct one bit but cannot distinguishes two-bit errors. To have two error bit detection, one extra parity is added to have SECDED, which has the minimal Hamming distance of four.

Table 6.2 Parity bit combination for Hamming code

Data bit position		1	2	3	4	5	6	7	8	9	10	11
Parity bit	1	x	x		x	x		x		x		x
(Hamming)	2	x		x	x		x	x			x	x
	3		x	x	x				x	x	x	x
	4					x	x	x	x	x	x	x
Extended parity bit (SECDED)		x	x	x	x	x	x	x	x	x	x	x

Since SECDED can correct one and detect two error bits, it is commonly used in computing systems such as ECC-DRAM. Several variations of SECDED that can correct and detect adjacent error bits have also been proposed. These types of codes can obtain such features by specifically designing the parity bit patterns and decoding method. There are several coding methods such as CRC (cyclic redundancy code) and Reed-Solomon for detection-only coding techniques.

In summary, information redundancy can help detect and correct error bits in the system. In addition, this type of method can be used for protecting the data integrity in memory, caches, or on-chip communication.

6.2.3   Software Approach

Several literature have covered software approaches for tolerating faulty systems. In short, it can be *algorithm-based fault tolerance* where the computation includes the correction method itself. *Testing and validating* the system is also an important aspect in this type of approach.

One of the most important methods in *software* approach is *check-pointing* and *roll-back*. A snapshot (check-point) of the system will be stored periodically or due to some events. During the operation of the system, if the system is crashed, it tries to *roll-back* to one of the snapshots.

6.3   Fault-Tolerance for Neuromorphic Computing

This section summarizes the related works on protecting neuromorphic systems in three significant aspects: communication, computation, and memory. Furthermore, mapping methods for SNN to recover from faults are also summarized.

6.3.1   Memory Protection

Since memories are vulnerable to permanent and transient faults, protecting them is needed for highly reliable systems. One of the most popular methods is to use Error Correction Codes, such as Hamming or its extended version [14], which can correct one flipped bit in the *codeword*. For multiple bits upset, multi-bit correction such as Orthogonal Latin Square Code [15] or Triple Adjacent Error Correction [26] can be used. Another recovery method for memory is to add a spare row or column and use the spare one as a replacement for the faulty one [18].

On the other hand, memory errors can be tolerated in neural network applications by accepting a specific loss of accuracy. For example, as analyzed in [10], a CNN application lost 5.7% in terms of accuracy with an error rate of 0.0065. Furthermore, our analysis in Fig. 6.1 also shows an acceptable loss while inserting a similar error rate. In summary, we can either protect the memory using error correction code or accept accuracy loss under a certain noise level.

6.3.2   Communication Protection

Since spikes, neuron parameters, or weights could be transmitted within the system or external memories, corruption in these values could lead to inaccurate results. Therefore, protecting their integrity is crucial. Apparently, inheriting Error Correction Codes [14, 15, 26] from memory protection could be helpful. Here, the data is protected under a certain number of flipped bits.

Another type of error in communication is misrouting or arbitration failures [27]. In these cases, recovery using an alternative routing path or redundancy could be used. By avoiding the failure point and providing a viable routing, a fault-tolerant routing algorithm [4, 31] can help overcome these types of errors. On the other hand, by providing redundant modules [7], the system can replace a faulty module with a healthy one for recovery.

6.3.3   Computation Protection

Faults in a computation module could be critical to SNNs, as we previously demonstrated. Therefore, protecting computation units is substantially vital. In [32], the authors proposed a method to protect the systolic array by bypassing and retraining. By pruning the faulty part of computation and retraining the model, the system can accept a certain fault level. Johnson et al. [17] also presented a method to re-tune the spiking model with variable thresholds and operating frequencies to enable fault tolerance. A traditional method such as N-modular redundancies with a majority voting [23] could also be used to ensure the correctness in this case. However, it leads to high area costs and power consumption.

As large-scale SNN systems usually utilize Network-on-Chip as the communication infrastructure, the computation protection method can use spare cores and remapping algorithms. The fault-tolerance NoC system with homogeneous cores can be solved using Integer Linear Programming (ILP) as in [5], where the authors tried to optimize the communication cost (summary of the traveling distances). However, the ILP problem is NP-complete, which cannot deal with larger scales. The works in [5, 28] also present a Particle Swarm Optimisation (PSO) solution to reduce the complexity of the mapping algorithm. However, although the PSO-based approaches can significantly reduce the runtime, they cannot guarantee the optimized result.

Moreover, PSO has a high space complexity for storing all particles. For reliability aware mapping, Namazi et al. [25] presented an approach to map tasks to homogeneous NoC architecture using a Mixed Non-Linear Programming model. Despite providing promising results, the mentioned approaches only target conventional multi-core systems. For our large-scale SNN system, since each node can have multiple computing units itself, internal node recovery is also possible instead of requiring external spare cores. Also, the recovery methods do not take the migrating time between cores into account.

6.3.4   SNN Mapping for Tolerating Errors

One of the method to correct faulty neuromorphic system is to remap the neurons to avoid faulty ones. Since the targeted system is a NoC-based multi-cores system, we can use both the SNN mapping methods [1, 2, 16, 19, 22, 24] and conventional multi-core NoC mapping methods [5] for placing neurons. While the method in [5] optimizes the communication cost, work in [30] optimizes the traffic in on-chip network based neuromorphic systems. However, while the traditional multi-core mapping such as ILP or PSO proves their efficiency, mapping for NN is highly complicated due to many neurons. The conventional SNN mapping method has two phases [2]: (1) Partitioning: cluster the NN into groups of neurons; (2) Mapping: map the groups of neurons to hardware. However, there are some problems: (1) both graph partitioning and mapping are NP-hard, which might not be solved optimally in polynomial time; therefore, a non-optimal solution can be justified; (2)the layered SNN applications have the node as the layers itself; and (3) the conventional methods do not take into account the multi-casting manner in communication. Lagrange multipliers [19] can reduce the runtime complexity. However, we still observe the long execution time. Since mapping for each neuron is not feasible, partitioning then is a potential approach [22]. However, we have to note that partitioning is an NP-hard problem. In [2], the authors adopted the Kernighan-Lin (KL) partitioning method for reducing the complexity despite not providing optimal results.

6.3.5 Fault-Tolerant Remapping for Neuromorphic Computing

In this section, we first formulate the problem of remapping in a faulty neuromorphic system. We then present the proposed Algorithm for migrating the unmapped neurons.

6.3.5.1 Problem Formulation

Here, we assume that the working system S has N nodes (or neuron clusters) where each node has E_i ($i = 0, 1, \ldots, N - 1$) neurons. In serial systems [1, 12], the number of neurons is equivalent to the number of memory slots a node can stores. Note that the value of E can vary up to design and can be different between clusters in heterogeneous systems. In short, the total number of neurons in the system S is $X = \sum_{i=0}^{N-1} E_i$. Here, we also assume that the desired SNN application requires W neurons. A feasible application must have $W \leq X$.

Fault-Tolerance

Because we support fault-tolerance in our system architecture, we consider R spare neurons, which $R = X - W$ as the repairing source. Once $k \leq R$ neurons are faulty and must be removed from the system, our problem formulation is to remap these k neurons to R spare neurons. If $k > R$, the system S cannot correct, and an off-chip migration should be considered (i.e., plugging a new chip and migrate to it). If the system can remap the function of k neurons, parameters, and weight of k neurons, we archive k−fault tolerance. In term of repair-ability, we divide it into two levels:

- *Node-level recovery*: If the node has enough spare neurons to correct its failed ones, it corrects internally by remapping. If it fails, the *system-level recovery* is used. A copy of the weights and parameters stored externally is read and written to the spare neuron.
- *System-level recovery*: If there are not enough spare neurons in a node for its internal recovery, the migration of neurons happens across nodes of the system. A migrating neuron can move from its original node to a new one. The corresponding weight, mapping LUT elements, and neuron status are copied to the new neuron. If a node happens to have more neurons to be mapped than E_i as designed, the unmapped neurons will migrate.

In *system-level recovery*, if the communication is guaranteed as reliable, the system must support up to R fault-tolerance ($k = R$). Figure 6.6 shows the system model of S with $N = 9$ nodes of $E_i = 256$ neurons ($X = 2, 304$) and a possible solution. The system requires $W = 2000$ neurons to perform the application and maps the neuron uniformly, as shown in Fig. 6.6a. There are 33 or 34 spares neurons per node in this mapping example, which allows the node to correct up to 33 or 34 fault neurons. Figure 6.6b illustrates the case of node (0,0) has 10 defective neurons and they are internally corrected using *node-level recovery*. However, Fig. 6.6c shows the case of 100 defective neurons, which the node (0, 0) fails to recover using *node-level* repair.

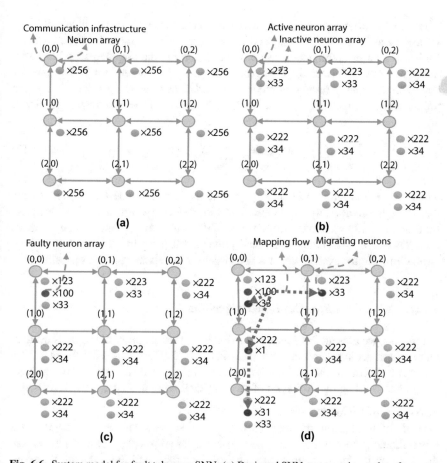

Fig. 6.6 System model for fault tolerance SNN: (**a**) Designed SNN system using nodes of neurons with an initial mapping; (**b**) *node-level recovery*; (**c**) The case *node-level recovery* fails to correct; (**d**) *System-level recovery*: a mapping flow of 100 faulty neurons to its node's neighbors. Values next the circle indicate the number of neurons in the circle type (gray: healthy and utilized; gray: healthy and spared; red: faulty; purple: migrating)

Figure 6.6d shows a mapping flow that map the faulty neuron to the node (0,0), (0,1) and (1,0), which are the current node and its neighbors.

Remapping Problem

One of the major problems is how to remap the SNN to recover from faulty neurons. Traditionally, one of the optimization goals for remapping is to minimize the following communication cost [5]:

$$F_{cost} = \sum_{i=0, j=0}^{W} d_{ij} \times c_{ij} \qquad (6.11)$$

where d_{ij} and c_{ij} are the distance and the connection status between node i and j. If we use c_{ij} as binary (0/1), F_{cost} is the sum of traveling distance between neurons. With this kind of optimization, we only need to rerun the mapping algorithm with faulty neurons' information. However, at large-scale systems, migrating neurons require an enormous amount of memory access. Therefore, this work optimizes the migration cost, which is the cost of migrating neurons of the new mapping method:

$$M_{cost} = \sum_{i=0, j=0}^{W} d_{ij} \times m_{ij} \qquad (6.12)$$

where m_{ij} is the number of migrating neurons between node i and j. Since the data (weight memory, threshold, etc..) within the faulty neurons can be corrupted, the system should write back from its host CPU. The moving distance is d_{0j}, where node 0 is where the I/O module is usually attached. This optimization considers high Availability for the system where it needs as least as repairing time as possible.

6.3.5.2   Max-Flow Min-Cut Based Algorithm

In this part, we present the Algorithm to enhance the reliability of the SNN system. Our main target is to optimize the M_{cost} in Eq. 6.12. We first present the max-flow min-cut theorem for the optimal flow. Then, the augmented versions multi-layers design and tackling the limitation of the max-flow min-cut theorem are discussed.

Max-Flow Min-Cut Theorem
One of the most common methods to find a flow between sink and source in a graph model is to use the max-flow min-cut theorem to optimize it. Here, we use the same principle: the sources are the faulty neurons, and the sinks are the spare ones. However, the multi-sink multi-source problem is usually complicated and can be converted to the conventional one using a virtual sink and a virtual source.

One of the main reasons to choose the max-flow min-cut approach as the solution for the remapping problem is a good trade-off between efficiency and execution time (or memory footprint). Compared to a greedy search approach from faulty neurons to spare neurons (evaluated later), the max-flow min-cut method provides a better mapping distance by creating a flow (chain of migrating). Meanwhile, the max-flow min-cut complexity is smaller than meta-heuristic methods (i.e., Genetic Algorithm or Particle Swarm Optimization). Moreover, these meta-heuristic methods require a huge memory footprint, which might not be optimal for a low-cost host CPU. Remapping the whole system by reusing the mapping method is a viable solution; however, as we previously discussed, the migration cost can be high due to the tremendous amount of memory transactions needed.

Figure 6.7 shows the flow graph for the fault tolerance in Fig. 6.6c. To support moving neurons, we firstly build a virtual source and virtual sink for the flow graph. Then, the connection between the virtual source to the faulty node has a capacity

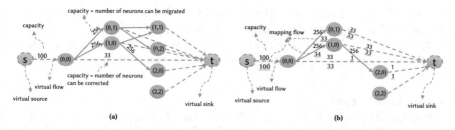

Fig. 6.7 Flow graph for max-flow min-cut problem: (**a**) Converted flow from the NoC-based SNN; (**b**) A solution of max-flow min-cut problem

as the number of faulty neurons (i.e., 100 in Fig. 6.7a). From the connected node, the flow capacity to its neighbors is the number of healthy neurons in the neighbor. For instance, Fig. 6.7a shows the capacity of the flow between (0,0) and (0,1) is 256 since the node (0,1) has 256 neurons and all can be migrated. For each node, there is a virtual flow to a virtual sink with the capacity of the number of spare neurons that are available to be used. For instance, Fig. 6.7a shows the capacity of the flow between (0,0) and t is 33 since the node (0,0) has 33 spare neurons. As we can realize in Fig. 6.7, the flow only comes from one node to one of its neighbors as we limit the traveling distance of migration to 1. In other words, Eq. 6.12 has $d_{ij} \leq 1$, which can reduce the migration cost. After solving using a max-flow min-cut solution, we end up having a flow map in Fig. 6.7b. Here, we can convert back to the NoC-based SNN to have the new mapping.

The max-flow min-cut theorem is applied as follows:

1. For all nodes creates a flow of migration between them. The capacity of the maximum number of neurons could be migrated via them.
2. Since we minimize the extra distance of migration, we use *maximum migrating distance* equal one ($d_{max} = 1; d_{ij} < d_{max}$). *Maximum migrating distance* (d_{max}) is the maximum number of hops that a neuron can migrate. With this $d_{max} = 1$ value, a neuron can only move to one of its four neighbors, which limits the capacity down to the maximum healthy number of neurons of the destination.
3. To allow neurons to migrate more than one hope, we can increase the *maximum migrating distance* value. The distance could constrain   item The movable distance of a neuron to its connected nodes.
4. Once we build all the nodes and the capacity of the flow between nodes.

As shown in Fig. 6.7, we know we can create a specific max-flow min-cut problem by making the flow graph. To solve this problem, we use the Edmonds–Karp algorithm to implement the Ford-Fulkerson method.

The Edmonds–Karp Algorithm has the run time complexity of $O(|V||E|^2)$ (E: number of edges, V: number of vertices), which can be translated to $O(N^3)$ for both 2D and 3D network (N: number of nodes). Therefore, the complexity of our mapping is guaranteed as P instead of NP. Meanwhile, heuristic search complexity is $O(N!)$, and ILP is NP-complete. On the other hand, PSO-based approach [28] has the complexity of $O(GKN^2 logN)$ (G: number of generation, K: number of particles). Since the number of generations or particles scale with the number of nodes, the PSO approach has a higher complexity than ours (PSO: $O(N^4 log N)$ if G and K scales linearly to N; ours: $O(N^3)$). However, the PSO approach [28] requires a massive number of particles, making it has a larger memory footprint than ours.

Graph-Based Algorithm

Algorithm 2 shows our proposed Algorithm for tolerating defective neurons. At first, it built the flow graph from sources and sink in lines 2–7. Then, for each node n_i, it adds an edge from the source with the capacity of the number of faulty neurons. As the flow goes out of the source, the Algorithm tries to fill the capacity as much as possible. We also connect the node n_i with a virtual sink with the number of spare neurons' capacity. At the end of this part, we built the flow ready for the *node-level recovery*.

Algorithm 2 The proposed max-flow min-cut neuron cluster replacement algorithm

---

```
// Build flow graph
```
1   add source s and sink t **for** *(node n_i in the system)* **do**
2   add vertex for the node n_i
   add edge from the source s to the vertex n_i
   add capacity $n_i \rightarrow s$ = number of defective neuron in the vertex n_i
   add edge from the vertex n_i to the sink t
   add capacity $n_i \rightarrow t$ = number of available redundant neurons attached to vertex n_i

3   **for** *(node n_i in the system)* **do**
   `// Node-level repair`
4   **if** *node n_i has more redundancies than defects* **then**
5   *node-level recovery* vertex n_i;

6   **for** *(node n_i in the system)* **do**
7   **for** *(node n_j in the system)* **do**
8   **if** *($d_{ij} \leq d_{max}$)* **then**
9   add edge from the vertex n_i to the vertex n_j
       add capacity $n_i \rightarrow n_j$ = number of healthy neurons in n_j

```
// System-level repair with Edmonds-Karp
```
10   **while** *no augmenting path* **do**
11   Breadth first search to find minimum path
   Augmenting the found minimum path with capacity
   Save the flow
```
// Finish the algorithm and require re-training or maintaince if
   it failed
```
12   **if** *(max-flow == k)* **then**
   `// done`
13   return 0;
14   **else**
   `// The approach fails to correct.`
15   return 1;

In the second part of the Algorithm, we first repair the system with *node-level recovery*. Then, we build the flow between nodes by adding the flow from a node to a node within the maximum distance d_{max}. The capacity is the maximum flow between those nodes, which is the minimum value of healthy neurons of the destination. For instance, the node n_j has 120 healthy neurons; the maximum capacity it can gain is 120 since the system can only migrate at most 120 neurons to it.

After completing the flow graph, we perform the Edmonds–Karp Algorithm to find the maximum flow and each edge's corresponding flow between the nodes. This number in each edge indicates the number of migrated neurons. The flow of the edge between the nodes and sink is the recovery using spare neurons. After completing the process, we now compare the maximum flow with the number of defective neurons (k). If they are equal, it means the Algorithm successfully corrects all k faulty neurons. If they are not equal, it means the max-flow min-cut implementation fails to recover. We will discuss the problem and how we can improve the Algorithm in the next section.

Figure 6.8 illustrates how our algorithm works in a 3D-NoC based neuromorphic system of $N = 3 \times 4 \times 3 = 36$ nodes. Each node consists of 256 neurons, which makes the total number of available neurons $X = 9216$. Only 9060 neurons are mapped, which leaves 156 neurons as spares. The system encounters eight faulty nodes with 138 defective neuron cases where only the node (2,2,0) with four faulty and 4 square neurons can complete the recovery using only node-level recovery. By migrating the unmapped neurons to their neighboring nodes, it allows *system-level recovery*. The neighboring nodes now have some unmapped neurons and look for the new nearby nodes. At the end of the Algorithm, it successfully maps the 138 faulty neurons, and there are 18 spare neurons left in the system. As shown in Fig. 6.8, the maximum movement of a neuron is only one hop from its original one. For instance, 7 neurons are migrated from (0,0,1) to (0,0,0). The node (0,0,0) also receives four neurons from (1,0,0), which leads to 11 neurons to map. By mapping 11 neurons and having four spares, the node (0,0,0) has seven unmapped and original neurons and migrates them to (0,1,0). By creating chains of migrations within the system, the proposed Algorithm helps recover the faulty neuron and minimize the traveling distance of an unmapped neuron's original node to the new node.

Augmenting Migrating Distance

Although using the max-flow min-cut method can optimize the neuron migration cost, the max-flow min-cut process is not optimal. To increase the minimum cut, we should relax the value of d_{max}; however, it increases the Edmonds–Karp Algorithm's complexity (increase number of edges). Based on the max-flow min-cut theorem, the maximum neurons that can be corrected can be limited by the minimal cut of the flow network. Therefore, there is a chance that the system cannot correct as much as neurons as its number of redundancies.

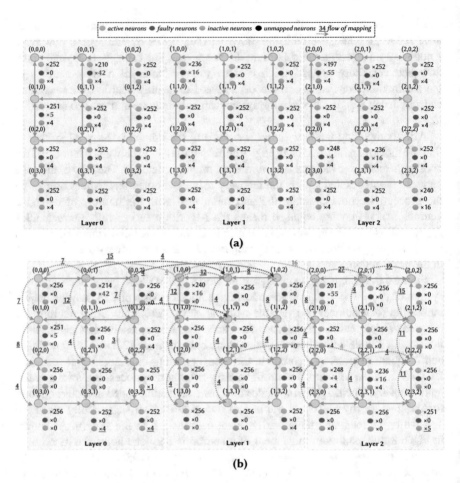

Fig. 6.8 An illustration of the proposed algorithm: (**a**) Faulty case; (**b**) Post-mapping using the proposed algorithm

Let us consider the node (0,0) or (0,0,0) in Fig. 6.9. These nodes have 200 faulty neurons and surrounded by nodes with 210 faulty neurons. After *node-level* recovery, there are 177 unmapped neurons. However, the maximum flow from these nodes to their neighbors is 112 and 168 for 2D and 3D mesh topology, respectively. In this case, the system fails to recover regardless of having redundancies in other nodes. In this fashion, we need to consider a communication cost of two, allowing neurons to move by two hops. The flow graph must be reconstructed for this different d_{max}.

Algorithm 3 shows our augmenting maximum migrating distance algorithm for tackling the problem of the small minimal cut section mentioned above. If the number of faults is larger than the number of spares, the mapping is not successful. Here, we need to run the Algorithm 2 depending on the single or multiple layers

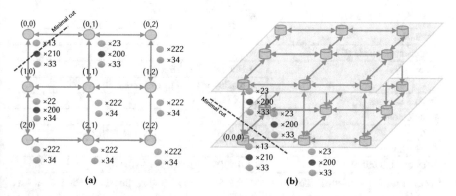

Fig. 6.9 Examples of the minimal cut drawback: (**a**) 2D-mesh; (**b**) 3D-mesh

---

Algorithm 3 Augmenting migrating distance algorithm

$d_{max} = 1$
while *(mapping success **or** d_{max} > (number of layers +number of rows + number of columns))* **do**
    run Algorithm 2
    $d_{max} + +$;

SNN. Later, we perform either retraining or maintenance for fitting the SNN to the neuromorphic system.

By gradually increasing the migrating distance, we can find the smallest value to recover the system failure. Therefore, we can balance the trade-off between the maximum migrating distance (d_{max}) and the recovery. Once the d_{max} value increases to a maximum distance within the NoC, we can ensure the mapping for the $k \leq R$ cases.

6.3.5.3   Evaluation

In this section, we evaluate the proposed algorithms (MFMC: *max-flow min-cut* adaption and GA: *Genetic Algorithm*), 1-hop and N-hop, and *Greedy Search* (GS) to understand their efficiency. The *Greedy Search* runs each node once and looks for a spare node within one (1) hop range or in the entire system (N-hop) with the shortest distance. The algorithms are implemented in Java. We insert the faults into the system to evaluate the efficiency of the algorithms. Here, we focus on the communication cost function F_{cost} in Eq. 6.11 and evaluate both 2D and 3D Meshes topology in terms of the migration efficiency. Different system sizes and fault rates are discussed. We evaluate the multi-layer perceptron (MLP) network. The MLP is organized in layers, and the neurons separated by one or more layers are not connected. The input spikes are feed to the router with the smallest indexes (i.e. (0,0) or (0,0,0)). In this evaluation, we measure two major parameters: *(1) mapping rate:* the ability to map the faulty neurons to the spare ones; *(2) average spike transmission cost (F_{cost}):* the average distance of all connections and *(3) Migration*

Table 6.3 Configuration for the evaluation[a]

Parameter	Value
# neurons per node (E)	256
# nodes (N)	2D-NoC: 4×4 to 16×16
	3D-NoC: $4 \times 4 \times 4$ to $16 \times 16 \times 16$
# spare neurons (R)	$0.2 \times X$
# spare node	1
# faults (k)	$0.05 \times X$, $0.10 \times X$, $0.15 \times X$, and $0.20 \times X$
SNN # layers	4
SNN configuration[a]	$784:0.5*(W-10): 0.5*(W-10): 10$

[a] MLP model for MNIST. For example, the SNN configuration for E=256 and 4×4 is 784:1633:1633:10

cost M_{cost}: *the amount of read/write neurons need to adapt the system.*. The configuration of the evaluation is shown in Table 6.3. Figures 6.10 and 6.11 illustrate the results for the proposed system for 2D-NoC and 3D-NoC configurations (see Table 6.3). As shown in Figs. 6.10 and 6.11, our methods can map all faulty neurons to the spare ones regardless of the size or topology. We have to note that the MFMC algorithm is not optimal for communication costs and 1-hop *Greedy Search* can only map around 60% (around 80% with the worst cases) of the faulty neurons. This is because 1-hop *Greedy Search* only runs for once and looks for one mapping solution of its neighbor to fail to map easily. Meanwhile, the N-hop *Greedy Search* and the Genetic Algorithm can map all neurons.

The average F_{cost} (communication cost) also varies between different approaches. Since the 1-hop GS mostly fails to map the neurons, the average communication distance per neuron is unchanged. For other methods, the average F_{cost} fluctuates between different sizes. However, as we can observe in Figs. 6.10 and 6.11, they are reduced when we increased the size of the NoC. This due to the fact when we increase the size of the NoC, the impact of moving neurons is reduced. The effects are also smaller, with smaller fault rates (k values). We can even notice the communication cost maintains with remapping; however, we can observe a slight reduction with the migration-based Algorithm. Also, GA seems to have a better average F_{cost} since it reduces that value as the second factor.

On the other hand, the M_{cost} of MFMC is better than both GA and GS in most cases. However, under [4, 4] and $f = 0.05$ instances, we observe that the M_{cost} of MFMC is worse than the GA. This phenomenon can be explained by the fact that the GA can provide an optimal result (globally or locally) once it converges. Meanwhile, MFMC only tries the maximize the flow between faulty neurons and spare ones. However, once we increase the network's size or change to 3D-NoC, *MigSpike* easily dominates GA and GS. Thus, while GS is not an optimal approach, GA might need adjustments to find the optimal solution (i.e., different evolving methods or more generations). However, as we will discuss in the execution time evaluation, GA costs a long time to execute, limiting its efficiency.

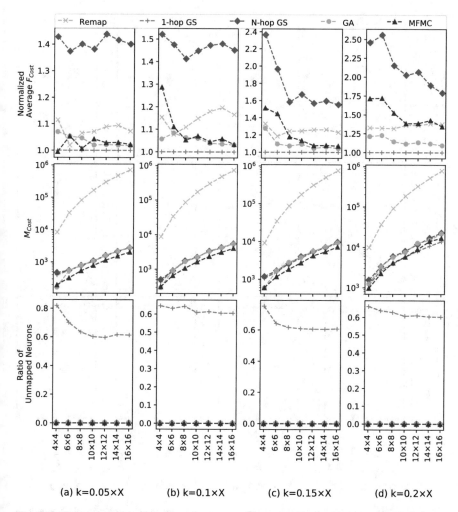

Fig. 6.10 Output mapping for migrated neurons with random fault patterns in 2D-NoCs. The system has 256 neurons per node; 20% of neurons are spare with 1 redundant node without any allocated neuron at 0% fault rate

As we presented in Fig. 6.9, one of the significant drawbacks of the max-flow min-cut method is the case where the minimal-cut is too small and creates the bottleneck. The groups' border can be recovered with MFMC; however, the central node cannot make it. With typical Ford-Fulkerson implementation, we can see that around 20% of the faulty node cannot be re-mapped, as shown as MFMC in Fig. 6.9. By relaxing the value of d_{max}, the MFMC-AMD system can map 100% of the faulty node (Figs. 6.12 and 6.13).

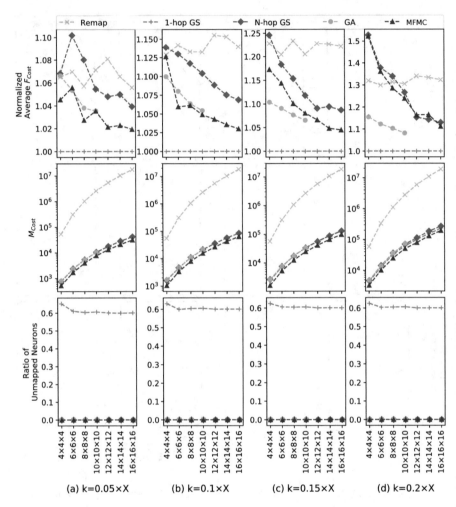

Fig. 6.11 Output mapping for migrated neurons with random fault patterns in 3D-NoCs. The system has 256 neurons per node; 20% of neurons are spare with 1 redundant node without any allocated neuron at 0% fault rate

6.4   Chapter Summary

In summary, this chapter has presented fault-tolerance features for neuromorphic systems. The types of fault and reliability measurement are firstly presented. Then, the chapter provides analyses on the impact of faults on neuromorphic systems. The conventional fault-tolerance methods for computing system are later presented. As a case study, this chapter shows a fault-tolerance design for NoC-based neuromorphic system using task migration and max-flow-min-cut theorem.

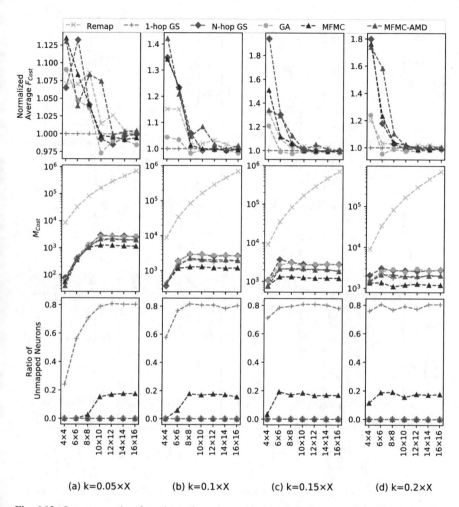

Fig. 6.12 Output mapping for migrated neurons with the minimal-cut cases in 2D NoCs. The system has 256 neurons per node; 20% of neurons are spare with 1 redundant node without any allocated neuron at 0% fault rate. MigSpike-AMD: the augmenting migrating distance d_{max} method

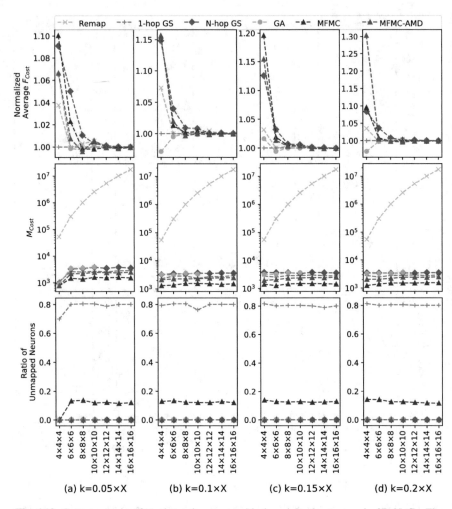

Fig. 6.13 Output mapping for migrated neurons with the minimal-cut cases in 3D NoCs. The system has 256 neurons per node; 20% of neurons are spare with 1 redundant node without any allocated neuron at 0% fault rate. MigSpike-AMD: the augmenting migrating distance d_{max} method

References

1. Akopyan F et al (2015) Truenorth: design and tool flow of a 65 mw 1 million neuron programmable neurosynaptic chip. IEEE Trans Comput-Aid Des Integr Circuits Syst 34(10):1537–1557
2. Balaji A. et al (2019) Mapping spiking neural networks to neuromorphic hardware. IEEE Trans Very Large Scale Integr Syst 28(1):76–86
3. Baumann R (2005) Soft errors in advanced computer systems. IEEE Des Test Comput 22(3):258–266
4. Ben Ahmed A, Ben Abdallah A (2016) Adaptive fault-tolerant architecture and routing algorithm for reliable many-core 3D-NoC systems. J Parallel Distrib Comput 93–94:30–43
5. Bhanu PV, Kulkarni PV, Soumya J (2019) Fault-tolerant network-on-chip design with flexible spare core placement. J Emerg Technol Comput Syst 15(1):1–23
6. Constantinescu C (2003) Trends and challenges in VLSI circuit reliability. IEEE Micro 23(4):14–19
7. Constantinides K, Plaza S, Blome J, Zhang B, Bertacco V, Mahlke S, Austin T, Orshansky M (2006) Bulletproof: adefect-tolerant CMP switch architecture. In: The twelfth international symposium on high-performance computer architecture. IEEE, Piscataway, pp 5–16
8. Dang KN, Tran XT (2018) Parity-based ECC and mechanism for detecting and correcting soft errors in on-chip communication. In: 2018 IEEE 12th international symposium on embedded multicore/many-core systems-on-chip (MCSoC).
9. Dang KN, Meyer M, Okuyama Y, Abdallah AB (2017) A low-overhead soft–hard fault-tolerant architecture, design and management scheme for reliable high-performance many-core 3D-NoC systems. J Supercomput 73(6):2705–2729
10. Denkinger BW, Ponzina F, Basu SS, Bonetti A, Balási S, Ruggiero M, Peón-Quirós M, Rossi D, Burg A, Atienza D (2019) Impact of memory voltage scaling on accuracy and resilience of deep learning based edge devices. IEEE Des Test 37:84–92
11. Diehl PU, Cook M (2015) Unsupervised learning of digit recognition using spike-timing-dependent plasticity. Front Comput Neurosci 9:99
12. Frenkel C et al (2018) A 0.086-mm^2 12.7-pj/sop 64k-synapse 256-neuron online-learning digital spiking neuromorphic processor in 28-nm CMOS. IEEE Trans Biomed Circuits Syst 13(1):145–158.
13. Hazan H et al (2018) BindsNET: a machine learning-oriented spiking neural networks library in Python. Front Neuroinf 12:89
14. Hsiao MY (1970) A class of optimal minimum odd-weight-column SEC-DED codes. IBM J Res Dev 14(4):395–401
15. Hsiao M, Bossen D, Chien R (1970) Orthogonal latin square codes. IBM J Res Dev 14(4):390–394
16. Jin X (2010) Parallel simulation of neural networks on spinnaker universal neuromorphic hardware. Ph.D. Thesis, The University of Manchester
17. Johnson AP, Liu J, Millard AG, Karim S, Tyrrell AM, Harkin J, Timmis J, McDaid LJ, Halliday DM (2017) Homeostatic fault tolerance in spiking neural networks: a dynamic hardware perspective. IEEE Trans Circuits Syst I Regul Pap 65(2):687–699
18. Kim I, Zorian Y, Komoriya G, Pham H, Higgins FP, Lewandowski JL (1998) Built in self repair for embedded high density SRAM. In: Proceedings international test conference 1998 (IEEE Cat. No. 98CH36270), pp 1112–1119
19. Kim G, Kornijcuk V, Kim J, Hwang CS, Jeong DS (2020) Optimal distribution of spiking neurons over multicore neuromorphic processors. IEEE Access 8:69426–69437
20. Koren I, Breuer MA (1984) On area and yield considerations for fault-tolerant VLSI processor arrays. IEEE Trans Comput 100(1):21–27
21. Koren I, Krishna CM (2020) Fault-Tolerant Systems. Morgan Kaufmann, Burlington

22. Li S, Guo S, Zhang L, Kang Z, Wang S, Shi W, Wang L, Xu W (2020) SNEAP: a fast and efficient toolchain for mapping large-scale spiking neural network onto NoC-based neuromorphic platform. arXiv:2004.01639
23. Lyons RE, Vanderkulk W (1962) The use of triple-modular redundancy to improve computer reliability. IBM J Res Dev 6(2):200–209
24. Moradi S, Qiao N, Stefanini F, Indiveri G (2017) A scalable multicore architecture with heterogeneous memory structures for dynamic neuromorphic asynchronous processors (DYNAPS). IEEE Trans Biomed Circuits Syst 12(1):106–122
25. Namazi A., Abdollahi M, Safari S, Mohammadi S (2017) A majority-based reliability-aware task mapping in high-performance homogenous NoC architectures. ACM Trans Embed Comput Syst 17(1):1–31
26. Neale A, Jonkman M, Sachdev M (2014) Adjacent-MBU-tolerant SEC-DED-TAEC-yAED codes for embedded SRAMS. IEEE Trans Circuits Syst II Exp Briefs 62(4):387–391
27. Prodromou A, Panteli A, Nicopoulos C, Sazeides Y (2012) Nocalert: an on-line and real-time fault detection mechanism for network-on-chip architectures. In: 2012 45th annual IEEE/ACM international symposium on microarchitecture. IEEE, Piscataway, pp 60–71
28. Sahu PK, Shah T, Manna K, Chattopadhyay S (2013) Application mapping onto mesh-based network-on-chip using discrete particle swarm optimization. IEEE Trans Very Large Scale Integr Syst 22(2):300–312
29. Sridharan V, Stearley J, DeBardeleben N, Blanchard S, Gurumurthi S (2013) Feng Shui of supercomputer memory positional effects in dram and SRAM faults. In: SC'13: proceedings of the international conference on high performance computing, networking, storage and analysis, IEEE, Piscataway, pp 1–11
30. Urgese G, Barchi F, Macii E, Acquaviva A (2016) Optimizing network traffic for spiking neural network simulations on densely interconnected many-core neuromorphic platforms. IEEE Trans Emerg Top Comput 6(3):317–329
31. Vu TH, Ikechukwu OM, Ben Abdallah A (2019) Fault-tolerant spike routing algorithm and architecture for three dimensional NoC-based neuromorphic systems. IEEE Access 7:90436–90452
32. Zhang JJ, Basu K, Garg S (2019) Fault-tolerant systolic array based accelerators for deep neural network execution. IEEE Des Test 36(5):44–53
33. Ziegler JF, Lanford WA (1981) The effect of sea level cosmic rays on electronic devices. J Appl Phys 52(6):4305–4312

Chapter 7
Reconfigurable Neuromorphic Computing System

Abstract The human brain can be characterized by its massive parallel reconfigurable synapses connecting billions of neurons. Synapses play a vital role in achieving the learning and adaptability of the human brain. The weight of a synapse shows connection strength between the two neurons linked by that synapse. Spiking neural networks are used in applications ranging from vision systems to brain-computer interfaces. However, the design of such systems has mainly focused on fixed functionality using available off-the-shelf components. Such an approach is lacking the flexibility to adapt to various computing environments. The reconfigurable design approach supports multiple target applications via dynamic reconfigurability, network topology independence, and network expandability. This chapter presents the architecture and hardware design of a reconfigurable neuromorphic processor. The architecture implements a spiking neural network that can be reconfigured to recover from faults with suitable methods that use an FPGA without being dependent on FPGA intellectual property. This approach makes possible its implementation in Application-Specific Integrated Circuits (ASICs).

7.1 Introduction

Neuromorphic systems have been used in applications ranging from vision systems [29] and brain-computer interface [54], to the simulation of the information processing in the biological brain [49]. Moreover, neuromorphic systems have allowed for the real-time processing of massive networks, which has proven valuable for neuro-robotics control and decision-making applications.

Simulation of the information processing of a biological brain requires interconnecting many parallel arrays of neurons. Unlike a multilayer perceptron neural network where all neurons fire at every propagation cycle, an SNN fires only when its voltage potential is stimulated beyond a threshold value [35].

When information is encoded as spikes, SNN employs a coding scheme which could be rate coding, population coding, or temporal coding [46]. Several spiking neuron models exist, and one of the prevalent ones often found in typical SNNs is the integrate and fire model [16]. The neuronal dynamics of this model are

Fig. 7.1 Typical DNN
accelerator organization

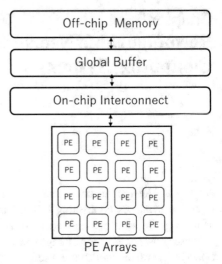

conceived as an integration process, together with a spiking mechanism. Typical spikes, irrespective of their amplitude and shape, are handled as similar events, and from the outset to finish, lasts about two milliseconds [4] traveling down axonal lengths. Another spiking neuron model noted for its detailed simulation of a biological neuron's ion channels is the Hodgkin and Huxley model [25]. This model is nonlinear and stochastic. However, it is complex, making it less ideal for large-scale simulation and hardware implementation. A typical artificial neural network (ANN) consists of several layers, and this has brought about the term deep neural network (DNN). Each layer in a DNN is often expressed as a 2D structure, making the network itself as a 3D structure. Mapping such a 3D structure onto a 2D circuit requires long wires between layers or congestion points (Fig. 7.1).

Several hurdles need to be surmounted to build in hardware a neuromorphic architecture with many synapses. The major problems that need to be surmounted are efficient on-chip communication and network routing, lightweight spiking neuro processing cores, on-chip learning, and an efficient neuro-coding scheme. Furthermore, we need to consider that the number of neurons to be connected are magnitudes of times (at least 10^3) larger than the number of cores that need to be interconnected on the recent multicore system on chip platforms [21]. These hurdles make the building of such a neuromorphic IC a challenging on-chip interconnect [11].

Recent progress in tract-tracing connectomics has helped deepen our understanding of the topology of the brain [3, 43] and has buttressed the findings that the anatomical topology of the brain network is organized as a three-dimensional small world network which is typified by dense local clustering of neurons with short connection lengths, and a few long-range connections between clusters [2, 51]. Therefore, the brain connectivity is generally described at three levels of scale: first, the single synaptic connections that link individual neurons at the micro-scale;

second, the networks connecting neuronal populations at the mesoscale; and third, brain regions linked by fiber pathways at the macro-scale [31]. In representing these connections in neuromorphic systems, a crossbar is one of the approaches employed. However, it has been recorded that the size of a crossbar directly affects the power consumption of a neuromorphic system, and this limitation in neuromorphic systems which employed single large crossbars [60, 62] have been demonstrated [1]. Therefore, for a scalable neuromorphic system that will support large SNN with a massive number of synapses to maintain low power, a partitioning and mapping of its synapses into smaller local crossbars linked using shared interconnect is a better approach. Sadly, with shared interconnect, the challenge of latency, which affects the timing of spikes, is introduced.

In SNN, the timing of synaptic spikes plays a vital role in the network's proper functioning. The timing of a postsynaptic spike is entirely influenced by the arrival time of presynaptic inputs [52]. Therefore any violation of this timing will negatively affect the operation of the spiking neurons. Also, there is a high level of local communication among neurons, so incoming spikes are distributed among neighboring neurons.

Various communication mediums are used when designing an interconnect, and they include shared bus and packet-switched Network on Chip (NoC) [65]. However, a shared bus is a poor choice when implementing a large-scale SNN with multicast routing since it suffers adversely with increased number of nodes. The nonlinear increase in neural connectivity will be too much for such an interconnect to handle. An interconnect that has been considered as a potential solution is the two-dimensional packet-switched network-on-chip (2D-NoC) [36]. However, this interconnects size increases with further scaling and begins to experience communication challenges that affect power and performance, especially in large-scale SNN chips. Three-dimensional packet-switched network-on-chip (3D-NoC), on the other hand, enables scaling and parallelism in the third dimension by combining NoC and 3D integrated circuits (3D-ICs) [6], and with the help of its short through-silicon-vias (TSVs) that enable communication between layers, it can reduce power. These merits of 3D-NoC make it suitable for large-scale SNN applications. It is to be noted that combining 3D ICs with multicore SoCs by stacking high-frequency cores will significantly increase the power and the thermals. Stacking 2D neural layers with low-frequency neuro-core will provide distributed parallel computations, which reduce power and the possibility of thermal hotspots.

In previous works [55, 58], a multicast 3D-NoC interconnect infrastructure for a neuromorphic system was proposed. Despite being known for having some underlying fault-tolerance attribute resulting from their densely parallel framework, SNNs face some fault challenges, especially those assumed from implementing them in hardware [53]. We proposed a fault-tolerant shortest path K-means-based multicast routing algorithm (FTSP-KMCR) to address challenges resulting from faulty links. In [42] we suggested a lightweight spiking neuron processing core suitable for the proposed 3D NoC.

7.2   Fault-Tolerant Neural Networks

There are several methods used to solve the fault occurrence problem in hardware implementations of neural networks. They are generally classified into learning, architecture, and hybrid-based approaches.

7.2.1   Learning-Based Approach

The learned-based methods are based on modified conventional learning rules for dealing with faults occurring in neural network systems. In [50], authors presented a fault-tolerant technique which temporarily injects faults in hidden neurons during training process. In this method, one to three neurons are randomly injected for each input example. Another work in [45] presented a modified training rule by adding a regularization term to the cost function. A work based on backpropagation was proposed in [59], for dealing with faults in classification tasks. In this learning method, weights are constrained under a limited range. In summary, although the modified learning methods do not require any external interactions afterward, they suffer a significant increase in the computation cost and take a long time for the training process.

Apart from the methods mentioned above, retraining methods are also wildly used. In [24], the authors proposed a technique that performs retraining periodically to improve fault-tolerance in GPGPUs systems. This method does not require either reprogramming or recompilation. The work in [17] proposed a retraining method for dealing with the impacts of timing errors in hardware-based neural networks. In this method, the retraining process is performed when timing errors influence output results. Authors in [39] presented a new learning rule mimicking the self-repair capability of the brain, in which the learning rule could reestablish the firing rate of neurons when synaptic faults occur.

7.2.2   Architecture-Based Approach

In architecture-based, the fault-tolerant methods are mainly based on the redundancy of the architecture. The redundancy is implemented in pre-trained networks, including hidden neurons and their connections. Work in [14] proposed a fault-tolerant architecture with the monotony of specific critical neurons. This reduces the hardware cost of the system. In this method, multiple sets of weight are stored in a processor, recomputing neural computations with multiple processors enables the system to detect and correct the faults in the processor, from there improving fault tolerance. Another work [20] also presented the redundancy of critical hidden neurons combined with a simple technique named augmentation. In the proposed method, the weight of the connections between augmented neurons and neurons in the output layer is half of its original value.

Apart from faulty neurons, faults in the connection between neurons have also been a concern. In dealing with faults occurring in connections and neurons, a method named weight shipping was proposed in [28]. When defects appear in some connections in this method, their weights are shifted to other fault-free connections of the same neuron. Besides, for a faulty neuron, its output connections are examined to be defective. A self-repairing hardware architecture was proposed in [34], as shown in Fig. 7.2. This architecture features self-detect and self-repair of synaptic faults and maintains the system performance with a fault rate of 40%. However, the experiment was taken with only two neurons, and the architecture may suffer a scalability limitation due to its area overhead. In SpiNNaker [61], an emergency routing was proposed to deal with congested or broken links in a

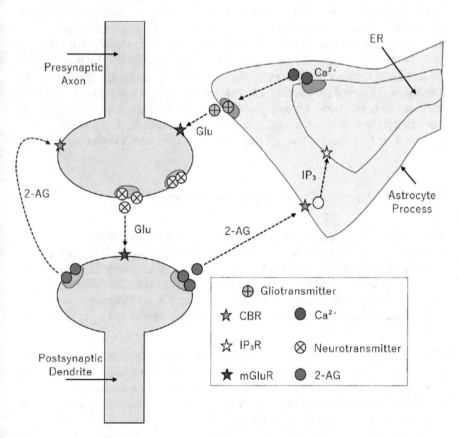

Fig. 7.2 A self-detect and self-repair mechanism mimicking capability in the human brain [34]. This mechanism is based on indirect feedback from the astrocyte cell (i.e., the most abundant type of glial cell in the brain), by regulating the synaptic transmission probability of release when faults occur

2D-NoC torus topology. The algorithm is based on redundancy in the NoC architecture to automatically redirect a blocked packet through adjacent links to its destination. This enables the system to avoid the timing violations of SNNs when congestion or faults occur.

7.2.3 Hybrid-Based Approach

Hybrid approaches are based on a combination of learning-based and architecture-based methods. In [44], a two-phase method was proposed to improve the fault-tolerance of a system. At the first phase, by feeding input and measuring the sensitivity, less important hidden neurons are removed. After that, some redundant neurons are added, the network is then retrained. The evaluation results show a fault-tolerant improvement of the system for two multiclass classification problems. This work was then extended in [13], where the authors proposed three methods: (1) during the backpropagation training, weights are restricted to have low magnitudes in order to avoid fault-tolerant degradation caused by high magnitude weights. To achieve the desired performance, hidden nodes are automatically added to the network. (2) During the training process, artificial faults are injected into some neurons and connections. (3) unimportant neurons are removed, while new neurons are added to share the role of critical neurons in the network. These methods were evaluated, and the results showed better robustness compared to other approaches.

7.3 Inter-Neuron Communication Network

Hardware implementations were proposed as alternative solutions to overcome the problems of the software simulation mentioned above. Such systems require high-parallelism and scalable interconnect architecture to convey huge number of spike generated from SNPCs. Hierarchical-bus, point-to-point, or NoC interconnect architectures are widely used, as illustrated in Fig. 7.3. In this section, we survey various interconnect platforms with spike routing methods for spiking neuromorphic systems.

Hierarchical Bus-Based Low-cost shared-bus based SNN architectures are proposed in [30, 38]. Although these approaches support multicast and broadcast routing, they suffer from scalability limitations when the network size increases. Other works were proposed in [9, 33]. These architectures boosted throughput; but, they were limited to small-size neural networks.

2D Packet-Switched-Based There are many ongoing SNN research projects based on 2D-NoC interconnects. Hereafter, we only review a few notable projects. The *Neurogrid* project [8] uses analog computation to emulate ion-channel activity and a digital communication scheme to support synaptic connections. The main

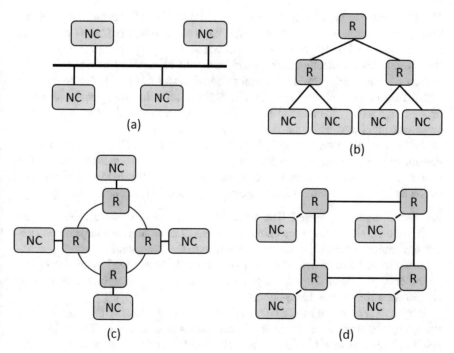

Fig. 7.3 Interconnect architectures for neuromorphic systems. (**a**) Shared bus. (**b**) Tree. (**c**) Ring.
(**d**) 2D Mesh

building block is the neuro-core, which can accommodate a total of 65,536 quadratic
integrate-and-fire neuron models, and it uses an external FPGA and bank of SRAMs
for digital communication between neighboring neuro-cores. The Neurogrid has a
limitation on the maximum number of neurons per layer (up to 2175 neurons) that
makes it unable to offer biological real-time behavior [8].

H-NoC [10] uses a hierarchical star-mesh topology to connect neurons. The
H-NoC is organized into three layers: module, tile, and cluster. At the bottom,
each module router can connect up to ten neural cells, each of them as a main
neural computation element that can host one or multiple neurons. In the same
fashion, ten module routers are connected to a tile router. An attractive work in [37]
proposed a combination of hierarchical architecture and mesh routing strategies.
The architecture consists of multiple levels of routers.

In SpiNNaker [22], the interconnection between each node is handled by a NoC
using six links, which is wrapped into a triangular lattice; this lattice is then folded
onto a surface of a toroid. A node comprises of a processor cores and two NoC
routers, in which one handles the communication between the microprocessors and
the peripherals, and the second controls the communications between processors
and neighbor nodes. *FACETS* [48] uses a mixed-signal and high-density hardware
neural network architecture based on a combination of analog neurons and a digital
multilayer bus communication scheme; all of them placed on an uncut wafer. The

FACETS hardware model consists of a large number of ASICs containing the analog neuron and synapse circuits. A full wafer can comprise 384 HICANN chips [48], resulting in a total of 196,608 neurons per wafer. To support the interconnection of the neuron, this architecture uses a combination of hierarchical buses for handling neuron communication inside the wafer and off-wafer routers implemented on an FPGA based on a 2D-torus topology. *FACETS* can offer hardware acceleration with up to $10\,\mu s$ inter-spike interval per wafer. However, the architecture consumes a large amount of power estimated at 1 kW per wafer [48].

Another work, named *ClosNN*, is presented in [26]. The *ClosNN* system uses a customized NoC architecture based on Clos topology for the neural network. It is designed with a high diameter of mesh and low bisection bandwidth of hierarchical tree. The architecture suffers from wire/router physical limitations.

3D Packet-Switched-Based The work in [5] investigated the architecture and design of a 3D stacked neuromorphic accelerator. The architecture targeted processing applications on a CMOS vision sensor next to the first neural network layer. The authors claimed that only modest adaptations would be required to use the system for other applications. The 3D stacking architecture used face-to-face bonding of two 20 cm wafers using micro-bumps.

Recent work was presented in [64] about a real-time digital neuromorphic system for the simulation of large-scale conductance-based SNNs. The architecture was implemented in six Altera Stratix III FPGA boards to simulate one million neurons [64]. An *AER* multicast routing mechanism was used for inter-neuron communications. Although the NoC architecture meets the requirements of the system, it is hardly deployed in embedded neuromorphic systems [19].

Apart from the works mentioned above, routing methods for NoC-based SNNs need to be taken into consideration. This is because the spike routing method affects the load balance across the network and also the spike latency. In general, these works can be classified as unicast-based [63], path-based [18], and tree-based [47]. A comparison between these methods is presented in [18]. The basic ideas of these algorithms are shown in Fig. 7.4. Compared to the others, unicast-based [63] is an easy way of implementing multicast with no hardware overhead. This is because a multicast package will be replicated at the source node and sent sequentially to destinations. However, this method leads to a large amount of traffic because of the injection of multiple copies. In path-based [18] approach, a routing path is established from the source and to each destination. Before sending, every packet header needs to contain a list of all destinations. Whenever the packet reaches a target, the information of that destination will be removed from the header. This helps the packet to be sequentially delivered to all destinations. A disadvantage of this method is that it requires a long time for the packet preparation at the source node. Besides, when increasing the size of destination sets (large size of SNNs), it is not efficient to implement because of the large header size of packets.

Drawbacks of path-based can be overcome by tree-based [47]. A "virtual" tree is constructed with the source node as the root and destinations as leaves in this approach. The packets are sent from the source, going along branches and reaching

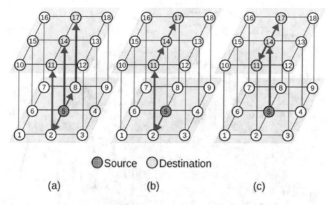

Fig. 7.4 Multicast routing mechanisms: (**a**) Unicast-based. (**b**) Path-based. (**c**) Tree-based

given destinations. Apart from advantages, a shortcoming of this method is high congestion in wormhole networks [32].

7.4 Reconfigurable Neuromorphic System Building Blocks

Figure 7.5 shows the architecture of a 3D NoC based neuromorphic architecture, named NASH, composed of several nodes, each of which is made up of a Spiking Neuron Processing Core (SNPC), a networkinterface (NI), and a fault-tolerant multicast 3D router (FTMC-3DR) connected in a 2D mesh topology, and stacked to form a 3D architecture. For design illustration, we show as an example 4x4 2D tiles stacked and interconnected with TSV to form a 3D architecture. Each SNPC embeds 256 LIF neurons with a crossbar-based synapse. The LIF model is adopted in NASH because it has proved to be effective for some learning applications and is suitable for digital implementation due to its modest hardware cost [27]. An output spike from a LIF neuron is sent to the postsynaptic neurons, either in the same SNPC or in another within the 3D network. If in the same SNPC, the postsynaptic neuron receives the spike and the weights of its synapse is obtained through the crossbar. But if the destination SNPC is different, the spike is encoded into a packet at the network interface and sent to the local FTMC-3DR, which routes it to the destination SNPC where the postsynaptic neuron resides. At the destination SNPC, the packet is decoded into a spike at the decoder, the postsynaptic neuron identified, and together with the synaptic weight obtained through the crossbar, it is sent to the postsynaptic neuron.

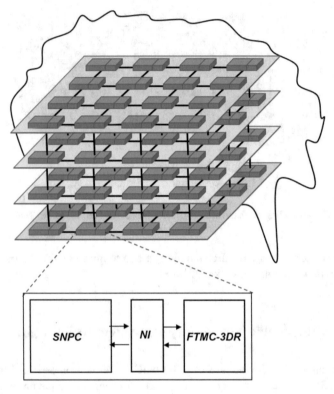

Fig. 7.5 Reconfigurable 3D-NoC based neuromorphic architecture

7.4.1 *Spiking Neuron Processing Core*

The spiking neuron processing core (SNPC) described in Fig. 7.6 is the processing element (PE) in the NASH system. It contains 256 physical leaky integrate and fire neurons, a crossbar-based synapse, a control unit, a synapse memory, an STDP learning block, and an encoder/decoder. The SNPC uses a spike array for spike events, and this was chosen in place of AER (Address Event Representation) to avoid memory overflow and extended pipeline time, which occurs when a large number of spikes are fired in the same time step. The SNPC design enables the state of neurons to be multiplexed onto a single shared bus of 256 bits, each bit taking the value of one or zero to signify the presence or absence of a spike event. The operation of the SNPC is controlled by the SNPC control unit Using seven states: *Idle, Dowload_spike, Generate_spike_&_Comp, Leak, Fire, Upload_spike*, and *Learn*. At the *Idle* state, the default state, the SNPC does nothing while waiting for input spikes. The arrival of an input spike train triggers the control unit to change its state to the *Dowload_spike* state to allow the input spike train to be received. After the spike train has been received, the next state,

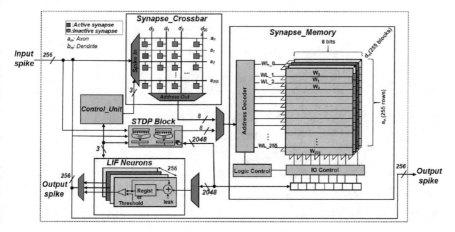

Fig. 7.6 Architecture of spiking neuron processing core (SNPC)

the *Generate_spike_&_Comp* state, is enabled. This state activates the crossbar to identify and update the destination neurons by activating their synapses using the addresses generated from the input spike train. After the last address has been updated, a signal is sent to the control unit to move to the *Leak* state, enabling a leak signal that causes a decay in the value of the membrane potential to be sent to the neurons. This state is followed by the *Fire* state, which activates the neurons to check for an output spike by comparing the membrane potential value with that of the set threshold. At the *Upload_spike* state, the spikes generated by the neurons are sent to the destination neurons.

7.4.1.1   LIF Neuron

A block diagram of the implemented LIF neuron model is described in Fig. 7.7. The neuron membrane potential is accumulated by adding up the input weighted spikes in the integrator, which uses extra 5-bits to handle the overflow. The resulting value is then stored in the 14-bit register, which utilizes 13 bits to keep the membrane potential value and 1 bit for overflow. To mimic the leak current found in the neural membrane, a set leak value that causes decay in the membrane potential value is received by the neuron at the end of the accumulation time step when the leak is activated. When the value of the accumulated membrane potential exceeds the threshold constant, an output spike which is represented by 1-bit is fired, and a signal is sent to the register to reset the value of the membrane potential to zero and start the refractory count, which gradually counts down every time step from the set refractory period to 0. Afterward, the neuron can accumulate incoming spikes again.

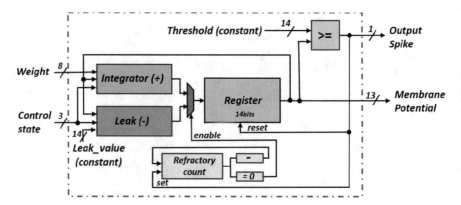

Fig. 7.7 LIF neuron architecture

7.4.1.2 Crossbar

The synapse crossbar architecture in the SNPC can be seen in Fig. 7.6. It represents the synaptic connections among neurons as intersections between axons and dendrites defined as horizontal and vertical wires arranged in an orthogonal manner. Each of the neurons has a fan in of 256, giving a total of 65k synapses for all embedded neurons. The weights of these synapses are stored on an on-chip SRAM. When the crossbar receives an input spike array, it is converted into synapse memory addresses via one hot process. The synaptic weights stored at these addresses are fetched and sent to the postsynaptic neurons. Leveraging the SNPC's architecture, the synapse crossbar can perform parallel neuron update, enabling the entire 256 neurons to be updated in one cycle.

7.4.1.3 Learning Algorithm

NASH implements on-chip learning based on the trace-based STDP Learning rule. The update logic of the implemented trace-based STDP is presented in Fig. 7.8. A single learning operation requires 16 presynaptic spike trace vectors, each from a simulation time step. To begin learning, a postsynaptic spike trace vector is verified. Then the presynaptic spike trace vectors are grouped into 8 *Before* and 8 *After* based on their arrival time relative to the postsynaptic spike trace vector. An OR operation is further performed on the *Before* spike vectors and on the *After* spike vectors to obtain two vectors. These two vectors are converted into synapse memory addresses, and Using the postsynaptic spike trace vector, the neurons whose synapses are to be updated are identified. The weights of these synapses are then fetched from synapse memory (SM) increased for the *Before* and decreased for the *After*, and then written back to synapse memory. The implemented trace-based STDP enables the parallel update of synapses.

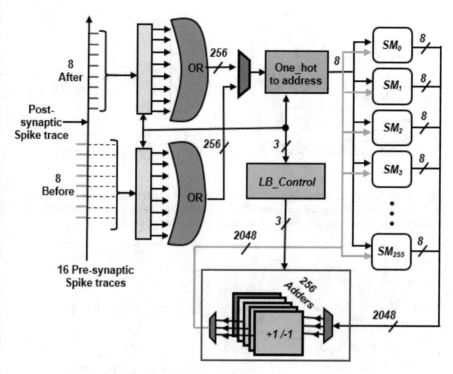

Fig. 7.8 STDP learning module architecture

7.4.2 Network Interface

The network interface (NI) consists of two modules: encoder, and decoder. The encoder serves as the output interface between an SNPC and its local FTMC-3DR. As presented in Fig. 7.9, the encoder is used to encode spikes into packets for transmission while the decoder is used to decode received packets to spikes. The 81 bits flit format is also shown in Fig. 7.9. The first two bits designated as Type indicate the type of flit it is; "00" for configuration and "11" for spike. The following 9 bits (3 bits for each X, Y, and Z dimension) are used to represent the address of the source neuron. The following 6 bits is a record of the time in which the source neuron fired the spike. The last 64 bits are used for the spike array.

7.4.3 Fault-Tolerant Multicast 3D Router

Figure 7.10 describes the architecture of the FTMC-3DR which is based on [7, 15]. It has 7 input(output) ports, one port for connecting to the local SNPC with which packets (spikes) can be injected into or received from the network, four for connecting to neighboring routers in the north, east, south, west, direction using the

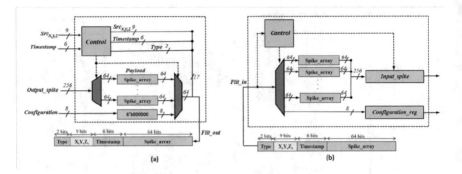

Fig. 7.9 Encoder and Decoder (Network interface to and from router). (**a**) Encoder encodes output spikes that will be transmitted from source SNPC to destination SNPCs into flits. (**b**) The Decoder on the other hand, decodes flits that arrive at a destination SNPC into spike

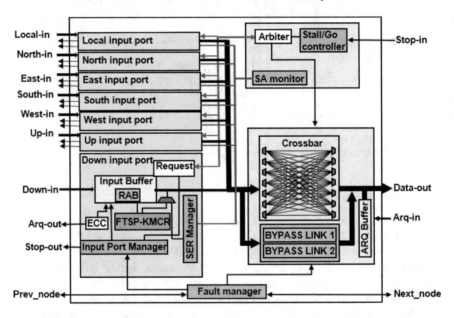

Fig. 7.10 Fault-tolerant multicast 3D router architecture

intralayer links, and the remaining up and down ports for those in the closest layers through TSVs. The switch-allocator (SA) and the crossbar control the transfer of packets (spikes) to the appropriate port. Each FTMC-3DR routes packets in four pipeline stages. In the first stage, buffer writing (BW), the received packets (spike) are stored in the port's input buffer. The second pipeline stage routing calculation (RC) obtains the source address of the stored packet from it and uses it to derive the next address, which is either in the X, Y, or Z dimension. After this address is derived, the switch-allocator, which handles the flow control (stall/go) and matrix arbitration (matrix-arbiter scheduler), is triggered in the third stage to allocate/make available the right port to the next router or local SNPC. After the right output port has been made available, the fourth stage, crossbar traversal (CT), begins, and the packet traverses the crossbar to the allocated output port.

7.5   Fault-Tolerant Spike Routing Algorithm

7.5.1   Shortest Path K-means Multicast Spike Routing Algorithm

Shortest Path K-means based MultiCast Routing algorithm (SP-KMCR) [57, 58] operates first by dividing destinations into subsets by adopting K-means from a given source, then the numbers of hops from the source to all the nodes in the subsets are calculated. For each subset, a node that has the shortest path to the source is detected (e. g., nodes *22* and *21* in Fig. 7.11a). The source sends its

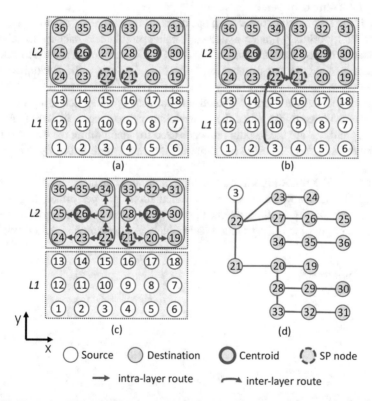

(a)

(b)

(c)

(d)

Source   Destination   Centroid   SP node

→ intra-layer route   ⌢ inter-layer route

Fig. 7.11 Example of SP-KMCR for a 6×3×2 3DNoC-SNN system, where nodes in *L1* send spike packets to all nodes in *L2*: (**a**) destinations are partitioned by adopting K-means clustering with centroids *26* and *29*, (**b**) the formation of the first part of the tree from a given source (node *3*) to shortest path node of each subgroup (SP node), (**c**) the second part of the tree from SP nodes to its destinations, (**d**) the routing tree from the given source to destinations

spike packets to the shortest path node of each subset to form the first part of the routing tree, therefore named SP-KMCR. The other part of the routing tree is formed from SP nodes to its destinations, as shown in Fig. 7.11c. Furthermore, it is worth mentioning that the SP-KMCR requires some computations for finding the shortest path. However, the computations are executed offline. The operation of the SP-KMCR is described in Algorithm 4

7.5.2 Fault-Tolerant K-means Multicast Spike Routing Algorithm

The shortest path fault-tolerant multicast routing algorithm is based on the SP-KMCR [55, 56]. The basic idea of the FTSP-KMCR is as follows: (1) offline computations of a primary routing tree from a given source node to its destinations and backup routing branches are performed. (2) After the offline calculation, the routing tables are configured.

The illustration of the primary and backup routing branches is shown in Fig. 7.12. When a faulty primary branch is detected, some pre-planned backup branch(es) is (are) used to bypass the faulty links. The SP-KMCR mechanism is used to calculate the branches (red) in the primary tree. On the other hand, the backup branches are alternative routes to the primary ones. For a considered router (i.e., "son"), the backup branches (green) are computed for the cases of faults occurring in primary connections. For example, when the father-to-son primary connection is faulty (i.e., pb_1), bb_1 and bb_2 are the backup branches used for maintaining the traffic between the "father" and "son". This is the same for the case where both pb_2 and pb_1 are faulty.

In the FTSP-KMCR algorithm, the computations of primary and backup routes are critical computational tasks. These calculations are performed offline. This allows to reduce the runtime overhead of the proposed routing algorithm; hence avoiding any possible timing violations in SNNs. As presented in Algorithm 5, the

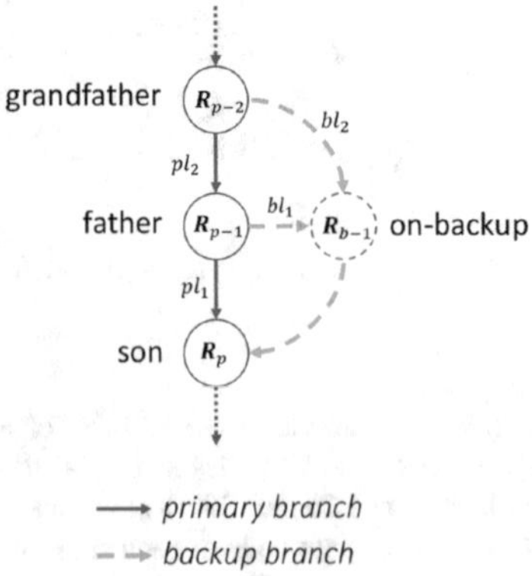

Fig. 7.12 Primary and backup branches

Algorithm 4 SP-KMCR multicast routing algorithm

/* *Input and output*   */
Input: // Source node address (S), destination node addresses (T), and the number of subsets (k)

16      $S = \{s_1(x_1, y_1, z_1), s_2(x_2, y_2, z_2), \ldots, s_n(x_n, y_n, z_n)\}$
        $T = \{t_1(x_1, y_1, z_1), t_2(x_2, y_2, z_2), \ldots, t_m(x_m, y_m, z_m)\}$
        k

Output: // Routing tree from each of source node to the destinations
17 **output** $P = \{p_1(s_1 \rightarrow T), p_2(s_2 \rightarrow T), \ldots p_n(s_n \rightarrow T)\}$

/* *Partition the destination set (T) into k subsets*   */
// Initial centroid nodes by randomly select from T
18 **foreach** $c_i \in C$ **do**
19 | $c_i \leftarrow t_j \in T$
20 **end**

// Evaluate centroid nodes and their labeled nodes
21 $C! = const$ // Calculate the distance between $t_i \in T$ to $c_j \in C$
22 **foreach** $t_i \in T$ **do**
23 | $d(t_i, c_j) = |x_i - x_j| + |y_i - y_j| + |z_i - z_j|$
24 **end**

// Assign each destination to its centroid by minimum distance
25 **foreach** $t_i \in T$ **do**
26 | $l(t_i) \leftarrow argmind(c_i, t_j)$
27 **end**
// Update centroid
28 **foreach** $c_i \in C$ **do**
29 | $c_i \leftarrow update(mean(t_{ij}))$
30 **end**

/* *Finding k shortest-path nodes (SP nodes) for every single*
   source node   */
31 **foreach** $s_i \in S$ **do**
32 | **foreach** $t_i \in T^k$ **do**
33 | | $d(s_i, t_j) = |x_i - x_j| + |y_i - y_j| + |z_i - z_j|$
34 | **end**
35 | $sp_i \leftarrow min(d(s_i, t_j))$
36 **end**

/* *Creating routing path from each source node to its SP nodes* */
37 **foreach** $s_i \in S$ **do**
38 | $p(s_i, sp_j) \leftarrow DOR\_based\_tree(s_i, sp_j)$
39 **end**

/* *Creating routing tree from each SP node to its destinations* */
40 **foreach** $sp_i \in SP$ **do**
41 | $p(sp_i, t_j) \leftarrow DOR\_based\_tree(sp_i, t_j)$
42 **end**

Algorithm 5 Off-line calculations of the primary and backup branches

---

/* **Input and output** */
Input: // Source node address (S), destination node addresses (T), and the number of subsets (k)

43      $S = \{s_1(x_1, y_1, z_1), s_2(x_2, y_2, z_2),\ldots,s_n(x_n, y_n, z_n)\}$
         $T = \{t_1(x_1, y_1, z_1), t_2(x_2, y_2, z_2),\ldots,t_m(x_m, y_m, z_m)\}$
         k

Output: //Primary (pr) and backup (bk) branches from S to T

44 **output** $P_{pr} = \{p_{pr,1}(s_1 \to T), p_{pr,2}(s_2 \to T), \ldots p_{pr,n}(s_n \to T)\}$ **output** $P_{bk} = \{p_{bk,1}(s_1 \to T), p_{bk,2}(s_2 \to T), \ldots p_{bk,n}(s_n \to T)\}$

/* **Centroid node assignment** */
// Initial centroid nodes by randomly select from T
45 **foreach** $c_i \in C$ **do**
46 | $c_i \leftarrow t_j \in T$
47 **end**

// Evaluate centroid nodes
48 (C != const) // Calculate the distance between $t_i \in T$ to $c_j \in C$
49 **foreach** $t_i \in T$ **do**
50 | $d(t_i, c_j) = |x_i - x_j| + |y_i - y_j| + |z_i - z_j|$
51 **end**

// Assign each destination to its centroid by minimum distance
52 **foreach** $t_i \in T$ **do**
53 | $l(t_i) \leftarrow argmind(c_i, t_j)$
54 **end**
// Update centroid
55 **foreach** $c_i \in C$ **do**
56 | $c_i \leftarrow update(mean(t_{ij}))$
57 **end**

/* **Finding the shortest paths** */
58 **foreach** $s_i \in S$ **do**
59 | **foreach** $t_i \in T^k$ **do**
60 | | $d(s_i, t_j) = |x_i - x_j| + |y_i - y_j| + |z_i - z_j|$
61 | **end**
62 | $sp_i \leftarrow min(d(s_i, t_j))$
63 **end**

/* **Creating primary and backup branches** */
// from each source to SP node
64 **foreach** $s_i \in S$ **do**
65 | $p_{pr}(s_i, sp_j) \leftarrow DOR^{v.1}\_based\_tree(s_i, sp_j)$
        $p_{bk}(s_i, sp_j) \leftarrow DOR^{v.\neq1}\_based\_tree(s_i, sp_j)$
66 **end**

// from each SP node to its destinations
67 **foreach** $sp_i \in SP$ **do**
68 | $p_{pr}(sp_i, t_j) \leftarrow DOR^{v.1}\_based\_tree(sp_i, t_j)$
        $p_{bk}(sp_i, t_j) \leftarrow DOR^{v.\neq1}\_based\_tree(sp_i, t_j)$
69 **end**

---

source and destination addresses (S, T) and the number of subsets (clusters) (k) are pre-defined as inputs, while output parts are a primary tree (P_{pr}) from each source to destinations and backup branches (P_{bk}). After that, the routing computation is done according to the following steps:

- **Step 1:** from destination addresses, destination subsets are determined by adopting k-means, as shown in lines *6–19*.
- **Step 2:** finding the shortest path from each source to a node (named $sp_i \in SP$) in each subset (with k subsets T^k, a given source node has k SP nodes), as depicted in lines *20–25*.
- **Step 3:** the first part of the primary tree is formed from the source node to SP ones. This is done by adopting dimension order routing (DOR) algorithm [12] from the source to each SP node, then merge with the same route. Alternative variations of the DOR are then adopted to calculate backup branches in order to guarantee that backup branches are separated from the primary routes. For example, if the formation of the primary tree uses DOR of ZYX, the backup branches use other variations of the DOR such as YZX or XZY.
- **Step 4:** following the same computation in step 2, the second part of the primary tree from SP nodes to their destinations in the same group and backup branches are calculated.

After the primary and backup routes are defined, they are used to configure the routing tables in routers. The pre-defined primary and backup routes are suitable for deploying SNN applications since the SNNs are pre-defined and mapped into the SNN system. Furthermore, this guarantees that the computation overhead of backup branches does not affect the proposed routing algorithm's recovery time and reduces the required hardware cost of the system. After the routing information is configured, the fault-management algorithm is implemented to handle incoming packets, as shown in Fig 7.13. For a given incoming packet, *fault_flag_val* is extracted to indicate whether the packet is in the primary or backup branch. At the same time, the source address is also used to compute its expected primary output port. In the case where *fault_flag_val* = 0 (i.e., the router plays the role of "father" or "grandfather"), the calculated *output_port* is then determined to be faulty or not. If it is not faulty, the packet is forwarded to the calculated output port in the primary branch. Otherwise, *output_port* is switched to use a backup_branch, and the *fault_flag_val* is also initiated to inform the next on-backup routers that this packet is on the backup branch. In the case where *fault_flag_val* $\neq$ 0 (i.e., the router role is as a on-backup or "son" router), the packet is routed through the output port as backup route, and *fault_flag_val* is also decreased by one.

7.6 Mapping

Mapping is one of the important areas of neuromorphic system design, especially those that employ multicore approach. The aim is to establish measurable links between parameters of the SNN application to be mapped, and those of the neuro-

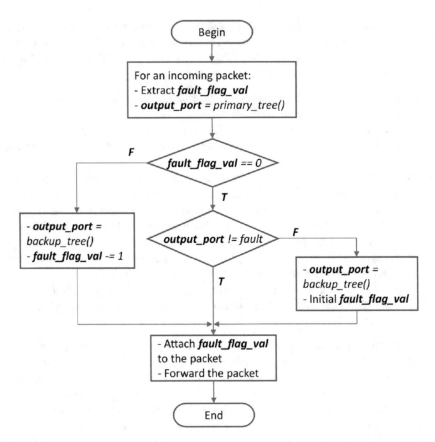

Fig. 7.13 Fault-management algorithm applied for "son", on-backup, "father" and "grandfather" routers

morphic system. The manner in which an SNN is mapped on a neuromorphic system significantly affects the performance and energy requirement of the neuromorphic system. As a result, an efficient mapping approach and method, is crucial to not just achieving good performance and energy efficiency, but also to provide a guide for efficient scaling of neuromorphic models. As illustrated in Fig. 7.14, an SNN of size 784:224:10 used for MNIST dataset classification is mapped on a 3 × 3 × 3 NASH configuration. This mapping approach is layer-based, where each network layer is mapped to a corresponding NASH layer. The input layer of 784 neurons is mapped to the first layer of NASH, utilizing 88 neurons from each of the 9 nodes in the layer. The hidden layer of 225 neurons is also mapped onto the second layer of NASH and utilized 25 neurons from each node in the layer. Finally, the output layer of 10 neurons is mapped to the third layer and utilizes five neurons each from two of the nodes in the layer.

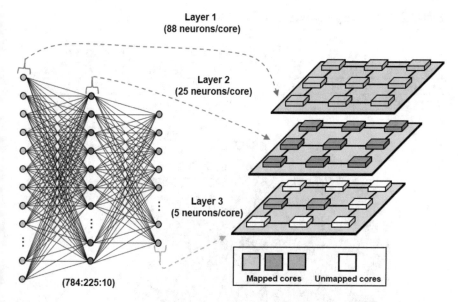

Fig. 7.14 784:225:10 SNN mapping on a 3 × 3 × 3 NASH configuration for MNIST classification application

The layer based mapping is sort of a naive approach to mapping. Several SNN applications may utilize a more complex network, leaving the layer based approach unfit to handle them. Therefore, a comprehensive and efficient mapping scheme is imperative.

7.7 Complexity Analysis

In analysing the complexity of NASH described in Sect. 7.4, the hardware design of the system was described in Verilog-HDL, and its synthesis and layout were made with Cadence tools. For ASIC implementation, the NANGATE 45 nm open-cell library [40] was used as the standard cells. OpenRAM [23] was used for generating the system memory, and TSV from FreePDK3D45 [41] was used for interlayer connection.

The design complexity of NASH and the baseline (2D) system nodes using the XY-UB and XYZ-UB algorithms, and also the SP-KMCR and FTSP-KMCR algorithms described in Sects. 7.5.1 and 7.5.2 respectively, are presented in Table 7.1. The design was made at a voltage of 1.1, and temperature of 25 °C, and from Table 7.1 we can see that the NASH node of these algorithms occupy a larger footprint and has higher power consumption when compared to the baseline system nodes. This is due to NASH's increased design complexity and its higher degree of path diversity enabled by TSVs, whose diameters also add to the footprint. An

Table 7.1 Design complexity comparison of NASH and the baseline nodes

	System					
	XY-UB	XYZ-UB	SP-KMCR		FTSP-KMCR	
Architecture	Baseline	NASH	Baseline	NASH	Baseline	NASH
Area (mm^2)	1.194	1.197	1.197	1.202	1.200	1.205
Power (mW)	49.29	50.02	49.58	50.10	50.86	51.76

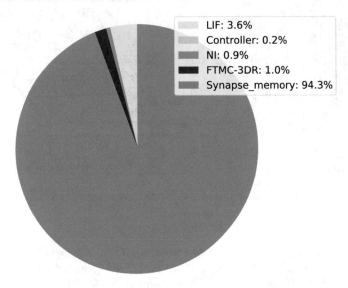

LIF: 3.6%
Controller: 0.2%
NI: 0.9%
FTMC-3DR: 1.0%
Synapse_memory: 94.3%

Fig. 7.15 Area analysis of NASH node

area analysis of the FTSP-KMCR NASH node which has an area of 1.205 mm^2 and consumes 51.76 mW is shown in Fig. 7.15. In this figure, we can see that over 94% of the node area is occupied by the synapse memory. This is because SNN is centered around synaptic operations. Therefore the amount of resources utilized in modeling these synapses takes up a significant portion of a neuromorphic system. Figure 7.16 illustrates the layout and the floorplan of the NASH system.

Using the MNIST application SNN and mapping approach described in Sect. 7.6, the accuracy of classifying 10K MNIST images on NASH using various synapse precision can be seen in Fig. 7.17. As presented in Table 7.2 with 8-bit synapse precision NASH reached an accuracy of 97.6%. However it can be seen that as the synapse precision decreases, so does the accuracy. The effect of varying synapse precision on design complexity can be seen in Fig. 7.18. With increased precision, the area and power complexity increases, and reduces otherwise.

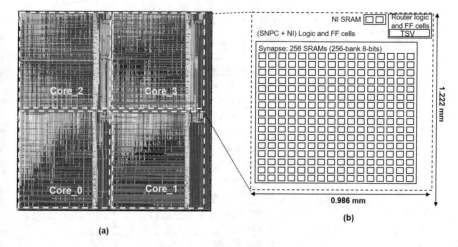

Fig. 7.16 (a) Layout of a 2 × 2 NASH layer. (b) A NASH node comprising of 256 neuron logic and 65k synapses in 256 SRAMs (256-bank 8-bits each), network interface logic and memory, and an FTMC-3DR logic and TSVs

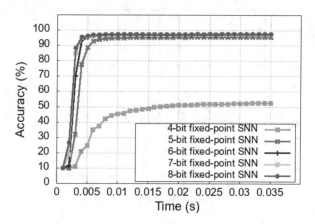

Fig. 7.17 Accuracy evaluation over various synapse precision

7.8 Chapter Summary

This chapter presented the architecture, hardware design, and complexity analysis of a reconfigurable neuromorphic system NASH, focusing on the SNPC, learning, interconnect, spike routing and mapping. The system leverages the high scalability and parallelism, low communication cost, and high throughput available in 3D-NoC to present a neuromorphic system capable of supporting large SNN with a massive number of synapses. The 256 physical neurons of the SNPC, enable parallel update of all neurons in a single time step. To handle the challenges that may arise in spike communication and lead to performance degradation, the FTSP-KMCR routing

Table 7.2 Summary of NASH node design complexity

Parameters/systems	This work
Benchmark	MNIST
Accuracy (%)	97.6
Number of cores	27
Neurons per core	256
Neuron model	LIF
Neuron update	Parallel
Synapses per core	65k
Synaptic connection	Crossbar
Synapse precision	8-bits
Learning rule	Off-chip SGD
Memory technology	SRAM
Interconnect	3D-NoC
Fault tolerance	Yes
Implementation	Digital
Technology node	45-nm NANGATE
Energy per synaptic operation (pJ)	11.3 pJ (1.1 V), 25 °C

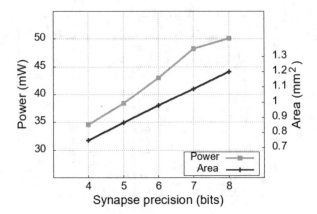

Fig. 7.18 Area and power evaluation over various synapse precision

algorithm was presented. The complexity analysis of NASH was also presented evaluating its area and power consumption with different routing algorithms and synapse precision, and its accuracy classifying MNIST images.

References

1. Balaji A, Das A, Wu Y, Huynh K, Dell'Anna F, Indiveri G, Krichmar JL, Dutt N, Schaafsma S, Catthoor F (2019) Mapping spiking neural networks to neuromorphic hardware
2. Bassett DS, Bullmore E (2006) Small-world brain networks. Neuroscientist 12(6):512–523

3. Bassett DS, Bullmore ET (2016) Small-world brain networks revisited. Neuroscientist 23(5):499–516

4. Bear M (2016) Neuroscience: exploring the brain. Wolters Kluwer, Philadelphia

5. Belhadj B, Valentian A, Vivet P, Duranton M, He L, Temam O (2014) The improbable but highly appropriate marriage of 3d stacking and neuromorphic accelerators. In: 2014 International conference on compilers, architecture and synthesis for embedded systems (CASES), Oct 2014, pp 1–9

6. Ben Abdallah A (2017) 3D integration technology for multicore systems on-chip. In: Advanced multicore systems-on-chip architecture, on-chip network, design. Springer, Singapore, pp 175–199

7. Ben Ahmed A, Ben Abdallah A (2016) Adaptive fault-tolerant architecture and routing algorithm for reliable many-core 3d-NoC systems. J Parallel Distrib Comput 93–94:30–43

8. Benjamin BV, Gao P, McQuinn E, Choudhary S, Chandrasekaran AR, Bussat J-M, Alvarez-Icaza R, Arthur JV, Merolla PA, Boahen K (2014) Neurogrid: a mixed-analog-digital multichip system for large-scale neural simulations. Proc IEEE 102(5):699–716

9. Boahen KA (1998) Communicating neuronal ensembles between neuromorphic chips. In: Neuromorphic systems engineering. Springer, Berlin, pp 229–259

10. Carrillo S (2013) Scalable hierarchical networks-on-chip architecture for brain-inspired computing. PhD thesis, University of Ulster

11. Carrillo S, Harkin J, McDaid LJ, Morgan F, Pande S, Cawley S, McGinley B (2013) Scalable hierarchical network-on-chip architecture for spiking neural network hardware implementations. IEEE Trans Parallel Distrib Syst 24(12):2451–2461

12. Chao C-H, Jheng K-Y, Wang H-Y, Wu J-C, Wu A-Y (2010) Traffic-and thermal-aware run-time thermal management scheme for 3d NoC systems. In: 2010 Fourth ACM/IEEE international symposium on networks-on-chip. IEEE, Piscataway, pp 223–230

13. Chin C-T, Mehrotra K, Mohan CK, Rankat S (1994) Training techniques to obtain fault-tolerant neural networks. In: Proceedings of IEEE 24th international symposium on fault-tolerant computing, June 1994, pp 360–369

14. Chu L, Wah BW (1990) Fault tolerant neural networks with hybrid redundancy. In: 1990 IJCNN international joint conference on neural networks, vol. 2, June 1990, pp 639–649

15. Dang KN, Ahmed AB, Okuyama Y, Abderazek BA (2017) Scalable design methodology and online algorithm for TSV-cluster defects recovery in highly reliable 3d-NoC systems. IEEE Trans Emerg Topics Comput 8(3):577–590

16. Dayan P, Abbott LF (2001) Theoretical neuroscience: computational and mathematical modeling of neural systems. The MIT Press, Cambridge

17. Deng J, Rang Y, Du Z, Wang Y, Li H, Temam O, Ienne P, Novo D, Li X, Chen Y, Wu C (2015) Retraining-based timing error mitigation for hardware neural networks. In: 2015 Design, automation test in Europe conference exhibition (DATE), March 2015, pp 593–596

18. Ebrahimi M (2014) Reliable and adaptive routing algorithms for 2d and 3d networks-on-chip. In: Routing algorithms in networks-on-chip. Springer, Berlin, pp 211–237

19. Ehsan MA, Zhou Z, Yi Y (2017) Modeling and analysis of neuronal membrane electrical activities in 3d neuromorphic computing system. In: 2017 IEEE international symposium on electromagnetic compatibility signal/power integrity (EMCSI), Aug 2017, pp 745–750

20. Emmerson MD, Damper RI (1993) Determining and improving the fault tolerance of multilayer perceptrons in a pattern-recognition application. IEEE Trans Neural Netw 4(5):788–793

21. Furber S, Temple S (2006) Neural systems engineering. J R Soc Interface 4(13):193–206

22. Furber SB, Galluppi F, Temple S, Plana LA (2014) The spinnaker project. Proc IEEE 102(5):652–665

23. Guthaus MR, Stine JE, Ataei S, Chen B, Wu B, Sarwar M (2016) Openram: an open-source memory compiler. In: 2016 IEEE/ACM international conference on computer-aided design (ICCAD), pp 1–6

24. Hashmi A, Berry H, Temam O, Lipasti M (2011) Automatic abstraction and fault tolerance in cortical microarchitectures. In: 2011 38th Annual international symposium on computer architecture (ISCA), June 2011, pp 1–10

25. Hodgkin AL, Huxley AF (1990) A quantitative description of membrane current and its application to conduction and excitation in nerve. Bull Math Biol 52(1):25–71
26. Hojabr R, Modarressi M, Daneshtalab M, Yasoubi A, Khonsari A (2017) Customizing clos network-on-chip for neural networks. IEEE Trans Comput 66(11):1865–1877
27. Indiveri G, Linares-Barranco B, Hamilton TJ, van Schaik A, Etienne-Cummings R, Delbruck T, Liu S-C, Dudek P, Häfliger P, Renaud S, Schemmel J, Cauwenberghs G, Arthur J, Hynna K, Folowosele F, Saighi S, Serrano-Gotarredona T, Wijekoon J, Wang Y, Boahen K (2011) Neuromorphic silicon neuron circuits. Front Neurosci 5:73
28. Khunasaraphan C, Vanapipat K, Lursinsap C (1994) Weight shifting techniques for self-recovery neural networks. IEEE Trans Neural Netw 5(4):651–658
29. Kulshrestha S (2016) Neuromorphic chips defence applications. SSRN Electronic J. https://doi.org/10.2139/ssrn.2773015
30. Lazzaro J, Wawrzynek J, Mahowald M, Sivilotti M, Gillespie D (1993) Silicon auditory processors as computer peripherals. IEEE Trans Neural Netw 4(3):523–528
31. Leergaard T, Hilgetag C, Sporns O (2012) Mapping the connectome: multi-level analysis of brain connectivity. Front Neuroinform 6:14
32. Lin X, Ni LM (1993) Multicast communication in multicomputer networks. IEEE Trans Parallel Distrib Syst 4(10):1105–1117
33. Liu S-C, Kramer J, Indiveri G, Delbrück T, Burg T, Douglas R (2001) Orientation-selective aVLSI spiking neurons. Neural Netw 14(6–7):629–643
34. Liu J, Harkin J, Maguire LP, McDaid LJ, Wade JJ (2018) Spanner: a self-repairing spiking neural network hardware architecture. IEEE Trans Neural Netw Learn Syst 29(4):1287–1300
35. Maass W (1997) Networks of spiking neurons: the third generation of neural network models. Neural Netw 10(9):1659–1671
36. Markram H, Gerstner W, Sjöström P (2012) Spike-timing-dependent plasticity: a comprehensive overview. Front Synaptic Neurosci 4:2
37. Moradi S, Manohar R (2018) The impact of on-chip communication on memory technologies for neuromorphic systems. J Phys D Appl Phys 52(1):014003
38. Mortara A, Vittoz EA, Venier P (1995) A communication scheme for analog VLSI perceptive systems. IEEE J Solid-State Circuits 30(6):660–669
39. Naeem M, McDaid LJ, Harkin J, Wade JJ, Marsland J (2015) On the role of astroglial syncytia in self-repairing spiking neural networks. IEEE Trans Neural Netw Learn Syst 26(10):2370–2380
40. NanGate Inc. (2014) Nangate open cell library 45 nm. http://www.nangate.com/. Accessed 16 June 2016
41. NCSU Electronic Design Automation (2015) FreePDK3D45 3D-IC process design kit. http://www.eda.ncsu.edu/wiki/FreePDK3D45:Contents. Accessed 16 June 2016
42. Ogbodo M, Vu T, Dang K, Abdallah A (2020) Light-weight spiking neuron processing core for large-scale 3d-NoC based spiking neural network processing systems. In: 2020 IEEE international conference on big data and smart computing (BigComp), pp 133–139
43. Ohno N, Katoh M, Saitoh Y, Saitoh S (2016) Recent advancement in the challenges to connectomics. Microscopy 65(2):97–107
44. Piche S (1992) Robustness of feedforward neural networks. In: [Proceedings 1992] IJCNN international joint conference on neural networks, June 1992, vol 2, pp 346–351
45. Poggio T, Girosi F (1990) Networks for approximation and learning. Proc IEEE 78(9):1481–1497
46. Rodrigues de Oliveira Neto J, Cerquinho Cajueiro JP, Ranhel J (2015) Neural encoding and spike generation for spiking neural networks implemented in FPGA. In: 2015 International conference on electronics, communications and computers (CONIELECOMP), pp 55–61

47. Samman F, Hollstein T, Glesner M (2010) New theory for deadlock-free multicast routing in wormhole-switched virtual-channelless networks-on-chip. IEEE Trans Parallel Distrib Syst 22(4):544–557
48. Schemmel J, Fieres J, Meier K (2008) Wafer-scale integration of analog neural networks. In: 2008 IEEE international joint conference on neural networks (IEEE world congress on computational intelligence). IEEE, Piscataway, pp 431–438
49. Sen-Bhattacharya B, James S, Rhodes O, Sugiarto I, Rowley A, Stokes AB, Gurney K, Furber SB (2018) Building a spiking neural network model of the basal ganglia on spinnaker. IEEE Trans Cogn Dev Syst 10(3):823–836
50. Sequin CH, Clay RD (1990) Fault tolerance in artificial neural networks. In: 1990 IJCNN international joint conference on neural networks, June 1990, vol 1, pp 703–708
51. Shibata S, Komaki Y, Seki F, Inouye MO, Nagai T, Okano H (2015) Connectomics: comprehensive approaches for whole-brain mapping. Microscopy 64(1):57–67
52. Taherkhani A, Belatreche A, Li Y, Maguire LP (2018) A supervised learning algorithm for learning precise timing of multiple spikes in multilayer spiking neural networks. IEEE Trans Neural Netw Learn Syst 29(11):5394–5407
53. Torres-Huitzil C, Girau B (2017) Fault and error tolerance in neural networks: a review. IEEE Access 5:17322–17341
54. Valencia D, Thies J, Alimohammad A (2019) Frameworks for efficient brain-computer interfacing. IEEE Trans Biomed Circuits Syst 13(6):1714–1722
55. Vu TH, Ikechukwu OM, Abdallah AB (2019) Fault-tolerant spike routing algorithm and architecture for three dimensional NoC-based neuromorphic systems. IEEE Access 7:90436–90452
56. Vu TH, Murakami Y, Abdallah AB (2019) Graceful fault-tolerant on-chip spike routing algorithm for mesh-based spiking neural networks. In: 2019 2nd International conference on intelligent autonomous systems (ICoIAS), Singapore, Feb 2019
57. Vu TH, Murakami Y, Abdallah AB (2019) A low-latency tree-based multicast spike routing for scalable multicore neuromorphic chips. In: ACM 5th International conference of computing for engineering and sciences, Hammamet, Tunisia, July 2019
58. Vu TH, Okuyama Y, Abdallah AB (2019) Comprehensive analytic performance assessment and k-means based multicast routing algorithm and architecture for 3d-NoC of spiking neurons. ACM J Emerg Technol Comput Syst 15(4):1–28
59. Wei N, Yang S, Tong S (1996) A modified learning algorithm for improving the fault tolerance of BP networks. In: Proceedings of international conference on neural networks (ICNN'96), June 1996, vol 1, pp 247–252
60. Wijesinghe P, Ankit A, Sengupta A, Roy K (2018) An all-memristor deep spiking neural computing system: a step towards realizing the low power, stochastic brain. IEEE Trans Emerg Topics Comput Intell 2(5), 345–358
61. Wu J, Furber S (2009) A multicast routing scheme for a universal spiking neural network architecture. Comput J 53(3):280–288
62. Xia Q, Yang JJ (2019) Memristive crossbar arrays for brain-inspired computing. Nat Mater 18(4):309–323
63. Xiang D, Shen K (2016) A new unicast-based multicast scheme for network-on-chip router and interconnect testing. ACM Trans Des Autom Electron Syst 21(2):1–23
64. Yang S, Wang J, Deng B, Liu C, Li H, Fietkiewicz C, Loparo KA (2018) Real-time neuromorphic system for large-scale conductance-based spiking neural networks. IEEE Trans Cybern 49(7), 2490–2503
65. Young AR, Dean ME, Plank JS, Rose GS (2019) A review of spiking neuromorphic hardware communication systems. IEEE Access 7:135606–135620

Chapter 8
Case Study: Real Hardware-Software Design of 3D-NoC-Based Neuromorphic System

Abstract This chapter presents the design and evaluation of a reliable three-dimensional digital neuromorphic processor (R-NASH) geared explicitly toward the 3D-ICs biological brain's three-dimensional structure. The platform enables high integration density and slight spike delay of spiking networks and features a scalable design. R-NASH is a design based on the Through-Silicon-Via technology, facilitating spiking neural network implementation on clustered neurons based on Network-on-Chip. In addition, we provide a memory interface with the host CPU, allowing for online training and inference of spiking neural networks. Moreover, R-NASH supports fault detection and recovery with graceful performance degradation.

8.1 Introduction

In spiking neural networks (SNNs), information is encoded using various encoding schemes, such as coincidence coding, rate coding, or temporal coding [21]. In addition, SNN typically employs an integrate-and-fire neurons model. A neuron generates voltage spikes (roughly 1 ms in duration per spike) that can travel down nerve fibers if they receive enough stimuli from other neurons with the presence of external stimuli. These pulses may vary in amplitude, shape, and duration, but they are generally treated as similar events. To better model the dynamics of the ion channel in a biological neuron, which is nonlinear and stochastic, the Hodgkin-Huxley [15] conductance-based neuron is often used. However, the Hodgkin-Huxley model is too complicated to be used for a large-scale simulation or hardware implementation.

Software simulation of SNN [16, 29] is a flexible method for investigating the behavior of neuronal systems. On the other hand, specialized hardware architectures with multiple neuro-cores could exploit the parallelism inherent within neural networks to provide high processing speeds with low power, which make SNNs suitable for embedded neuromorphic devices and control applications [30]. In general, the neuromorphic hardware systems consist of multiple nodes (or clusters of neurons) connected via an on-chip communication infrastructure [2, 23]. Expansion using a multi-chip system and off-chip interconnects is also a viable solution for scaling up SNNs [2, 9]. In recent years, integrating many neurons on a single chip while providing efficient and accurate learning has been investigated [2, 6, 9, 14, 27].

The challenges that need to be solved toward designing an efficient neuromorphic system include building a small-size parallel and reconfigurable architecture with low-power consumption, an efficient neuro-coding scheme, and an on-chip learning capability. Moreover, since the number of neurons to be connected is at least 10^3 times larger than the amount of PEs (Processing Elements) that need to be interconnected on modern multicore/multiprocessor SoC platforms [13], the on-chip communication and routing network is another major challenge. In a modern deep neural network (DNN) design, one neural network layer is often a 2D structure. However, the "mimicked" network is generally a 3D structure. Therefore, mapping a 3D structure onto 2D circuits may result in multiple long wires between layers or congestion points [7]. The constraints mentioned above make the deployment of such a brain-like IC a challenging on-chip interconnect problem [30].

An event-driven neuromorphic system relies on the arrival of spikes (action potentials) train to compute [25]. Therefore, the arrival times of action potentials are critical to allow accurate and consistent outputs. Since the shared bus is no longer suitable for multicore systems and point-to-point interconnects cannot serve a high fanout wires [20], moving to a new on-chip communication paradigm with the ability to extend to multiple-chip interconnects is needed. One of the consensuses of state-of-the-art architecture is to utilize the parallelism and scalability of 2D Network-on-Chip (NoCs) [2, 9] and further extend it to multichip systems. In this approach, the neurons of the silicon brain are clustered into nodes that are attached to micro-routers.

From another hand, semiconductor development is confronting the end of Moore's Law, which no longer allows us to reduce the feature size as we reach the atomic scale. To get to the "More than Moore" goal [31], heterogeneous integration is a suitable approach to integrate more transistors in the same die. One of the popular approaches is to stack the conventional 2D wafers together to form a 3D-chip [3]. Another method is monolithic 3D-ICs that support multiple silicon layers based on small vias [24]. The Through-Silicon Vias (TSVs) or Monolithic Intertier Vias (MIVs) constitute one of the main interlayer communication mediums. The 3D-Network-on-Chip (3D-NoC) [4, 5] is also a promising approach that can further enhance the parallelism and scalability of multicore and neuromorphic systems. Figure 8.1 illustrates a potential mapping of an emulated silicon brain into 3D-ICs. However, despite bringing several benefits of lower power, smaller footprints, and low latency, integrating a neuromorphic system into 3D-ICs was not well investigated.

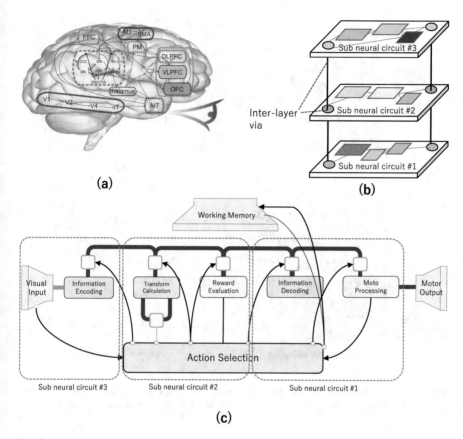

Fig. 8.1 Neuro-inspired 3D silicon brain. (**a**) The anatomical architecture of Spaun indicates major brain structures and their connectivity [12, 30]. (**b**) A possible mapping of a Spaun system into 3D-IC. (**c**) The architecture of Spaun, where thick dark-yellow lines illustrate communication between elements of the cortex while thin lines show connectivity between Basal Ganglia and the cortex

8.2   R-NASH System

The R-NASH platform design, shown in Fig. 8.2, consists of four phases. First, the software spiking neural network model *model* is developed with various configurable parameters, including neurons, synapses, thresholds, learning types, and interconnects. Then, the neurons mapping phase is performed based on four main steps: (1) *transformation*: convert the **mentioned** value to the binary that can be read and executed by the neuromorphic hardware, (2) *clusterize*: cluster and find the suitable mapping of neurons, (3) *scheduling*, and (4) *Configuration*. We use a genetic algorithm (discussed later) to optimize the flow. As a highly complex design, neuromorphic hardware is generally prone to soft and permanent faults, leading

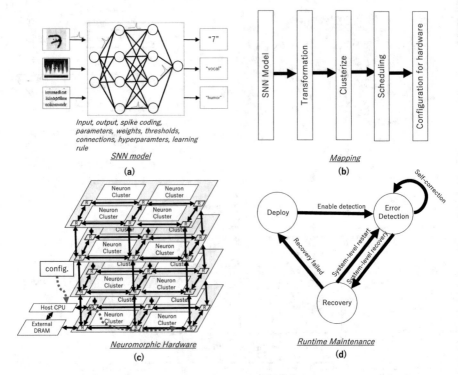

Fig. 8.2 R-NASH system platform. (**a**) SNN model. (**b**) Mapping into the neuromorphic hardware. (**c**) Our neuromorphic hardware in 3D-ICs. (**d**) Runtime maintenance for R-NASH

to performance degradation. Therefore, the *runtime maintenance* stage checks and recovers from faults.

8.3   R-NASH Hardware

8.3.1   R-NASH Hardware Building Blocks

The overall architecture of the neuromorphic system, shown in Fig. 8.3, is based on a 3D-IC approach to model the three-dimensional structure of the brain. The neurons and their synapses are clustered in neuron clusters or nodes. Instead of using a point-to-point neuron connection, we use a packet-switched mesh-based 3D-Network-on-Chip architecture. While a 3D-Mesh NoC handles the communication, the computation is done by neuron clusters (nodes) as shown in Fig. 8.3d. The incoming spikes in AER (Address-Event-Representation) protocol are stored in memory and decoded to obtain the address and the read enable signal for the weight memory. By reading the synapses from memory, the system emulates the weighted

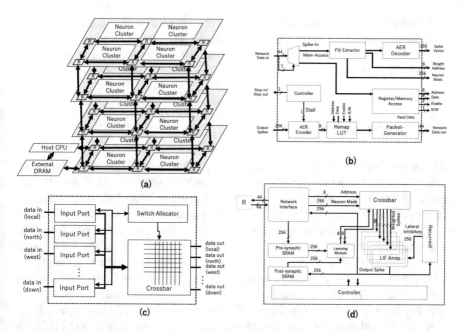

Fig. 8.3 R-NASH architecture. (**a**) Neuromorphic System. (**b**) Network-Interface. (**c**) 3D-Router. (**d**) SNPC core

spikes for LIF neuron inputs. After receiving the address of the corresponding synapse and the enable signal, the series of weighted inputs are sent to the dedicated LIF neuron, which accumulates the value, subtracts the leak, and checks the firing condition. The output spike is, finally, stored in a postsynaptic SRAM and sent to the Network-on-Chip. More details about the neuron clusters and the 3D NoC router architectures are given in Sects. 8.3.2 and 8.3.6, respectively.

8.3.2   Spiking Neural Processing Core (SNPC)

Figure 8.3d shows the architecture of the SNPC, which is the backbone of the R-NASH neuromorphic system. The SNPC consists of four major modules: (1) Network Interface, which supports the communication via 3D NoC; (2) Crossbar which help decode the spike and extract the corresponding weights; (3) LIF array, which consists of multiple LIF neurons performing in a parallel manner; and (4) Learning module which perform STDP learning. Once a packet is fed to NI from the 3D-NoC router, its information is decoded to decide the packet type. There are two types of packets: (1) Spikes (or action potentials) and (2) memory access (read/write). NI decodes to have the equivalent address in the memory and sends it to the crossbar for the spike flits under the AER format to obtain the weight. Once a neuron of the SNPC fires, its index or address will be encoded and sent to the 3D NoC to emulate inter-neuron communication. NI provides an interface to read and write each neuron's weight memory and parameters in single and burst transaction mode for memory access.

8.3.3   Network Interface

The Network Interface allows the neurons to communicate via the on-chip network infrastructure. R-NASH supports two types of flits: spike (in AER format) and memory flits. The AER format flit is converted to the address of the weight SRAM. A flit provides the instruction and the required addresses to read/write to/from the memory cells and registers in the neuron cluster. Here, the memory access flits are issued by a master (or external host) processor in the system. The NI supports two types of read/write commands: single and burst. The individual read/write only provides access to one element per request, while an argument of length must follow the burst ones. The NI converts the requested address to the local address of each weight memory or LIF array. Figure 8.4 illustrates R-NASH's two types of flit. The first bit indicates whether the flit is a spike (0) or memory access (1). With the spike flit, it is followed by four fields: (1) destination node address (9-bit), (2) neuron mask to allow the sparse connection (3-bit: 8 types of sparse), (3) AER of the source node (9-bit) and (4) AER of the neuron in the source node. Here, the AER of a firing neuron is represented via two files: node address and neuron address; this allows the system to scale up to $8 \times 8 \times 8$ 3D-NoC nodes (512 nodes) 256 neurons/node. We can extend it to 10-bit to allow 1024 neurons/node on the neuron ID, allowing the R-NASH system to have 0.5 million neurons and 0.5 billion synapses. To support a larger scale, we can extend all fields. For memory access spikes, there are four types: (1) single read, (2) burst read, (3) single write, and (4) burst write. The two-bit command field allows the system to inform the slave node to understand whether the transaction is done, kept, corrupted (need to rewrite/reread),

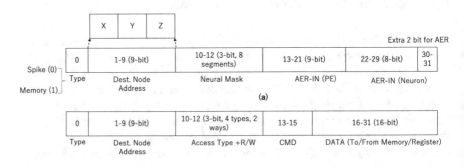

Fig. 8.4 R-NASH flit formats

or canceled. Since R-NASH is byte-addressable, the command field is followed by the address of memory/registers on the R-NASH node (16-bit). With the single read, the NI sends the data corresponding to the host node's address. With the burst read and write, the following flit consists of the length of these transactions. The details on how the node can access and provide an interface for the host CPU are shown in Sect. 8.3.3.

Figure 8.3b shows the block diagram of the Network Interface (NI). The input spikes are categorized into either input spikes or memory accesses. With the memory accesses, the NI provides an interface to read and write the data in all registers and memory blocks of the node. The read instruction makes the NI return the master processor value of the requested address. With the network's input spike, the NI decode phase gets the weight SRAM address and feeds it to the weight memory. For multi-layer SNNs or sparsity connections, the *Flit Extractor* provides the read enable (RE) signal for different layers or different links used in the weight memory. As a result, a node can have multiple AERs at the same address but for other neurons. The LIF array's output spike is fed into the AER decoder, which extracts the address of bit one (firing neuron). This address is then serially sent to the remap Look-Up-Table (LUT) to obtain the AER value of the receiving nodes.

8.3.4   Crossbar

As explained in the previous subsection, the input spikes (series of events determined by their timestamp and their polarity) are decoded to the weight address and neuron mask (read-enable signal) and fed to the crossbar. The crossbar is a set of SRAMs where each SRAM stores all synapses associated with a single neuron. The neuron mask signal is used to discard the unused weighted spike. After getting the address and the enable signal, the crossbar reads the synapses from memory, and sends them to the LIF array.

8.3.4.1   LIF Neuron

Due to its simplicity, the Leaky-Integrate-and-Fire (LIF) neuron model is selected for the R-NASH hardware architecture. Figure 8.5 shows the architecture of a LIF neuron. The weighted spike inputs (i_wspike) are fed into an adder and register structure for accumulation. At the end of each step, the leak's inverted value is also provided to the adder to reduce the membrane potential. The neuron firing condition is then validated by confirming that the membrane potential has exceeded the neuron firing threshold. After the neuron fires, it sets the refractory countdown and stops working until the countdown is over. The period of this countdown is the refractory period. Theoretically q, a LIF or IF computation is expressed with the equation

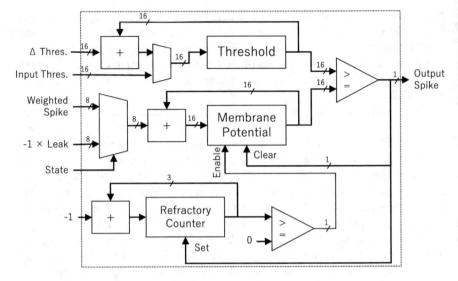

Fig. 8.5 LIF neuron architecture

below:

$$V_j(t) = V_j(t-1) + \sum_i w_{i,j} \times x_i(t-1) - \lambda \qquad (8.1)$$

Where, $V_j(t)$ is the membrane potential of neuron j at time step t, $w_{i,j}$ is the synapse weight between presynaptic neuron i and postsynaptic neuron j, $x_i(t-1)$ is the output spike of presynaptic neuron i, and λ is the leak constant ($\lambda = 0$ for IF). The output of the neuron j is described by the equation bellow:

$$x_j(t) = \begin{cases} 1, & \text{if } V_j(t) \geq Vt \\ 0, & \text{otherwise} \end{cases} \qquad (8.2)$$

At the crossbar, the input spikes are multiplied with the weights to have weighted inputs. These weighted inputs are accumulated as the membrane potential, and when the accumulated value exceeds the threshold, a spike is fired. The memory module handles the multiplication of the input spikes and the weight of synapses. Since the input spike is binary, there is no actual multiplication hardware. Therefore, a simple register and adder are used to perform the accumulation.

8.3.5   Controlling

There are two phases for controlling the R-NASH node: (1) training and (2) inference. If the training mode is enabled, the system enters the learning phases after each time step. If there is a postsynaptic spike of the emulated training time step, the weight is adjusted. Otherwise, the system skips to the next timestep. In the inference process, R-NASH starts with the synchronization of the timestep. Each node's registers are accessible via interfaces; therefore, the system can indicate, confirm, and change each node's timestep. Furthermore, since the definition of timestep is loosely defined, it helps the LIF array switch from integrating to the leaky phase. In other words, neurons can operate at different timesteps if needed. The operation of the SNPC follows four phases: (1) loading spikes, (2) integrating, (3) leak and firing, and (4) learning (optional). The first phase is for downloading spikes from presynaptic neurons via the interconnect. Due to the input cache's limited size, the spike is decoded and sent directly to the crossbar. On the other hand, it is packed to a spike vector for learning purposes due to its compactness in-memory footprint.

8.3.6   Inter-Neural Interconnect

R-NASH is based on a 3D-NoC that supports various fault-tolerance in input buffers, crossbar, routing hardware, and pipeline [7]. The inter-neural interconnect consists of multiple routers (R) to handle the communication between the neuron clusters [8]. Two types of flit are supported. The first type is the spike between neurons in AER format. The AER format flit is converted to the address of the weight SRAM. The second type of flit is memory access. To read and write to the memory cells and registers in the neuron cluster, a flit provides the instruction and the required argument (address). Here, the memory access flits are issued by a master (or external host) processor in the system. We support two types of read/write commands: single and burst. The individual read/write only provides access to one element per request, while a lengthy argument must follow the burst ones. The NI converts the requested address to the local address of each weight memory or LIF array. Figure 8.3b shows the block diagram of the Network Interface (NI). The input spikes are categorized into either input spikes or memory accesses. With the memory accesses, the NI provides an interface to read and write the data in all registers and memory blocks of the node. The read instruction makes the NI return the master processor value of the requested address. With the input spike from the network, the NI decode phase gets the weight SRAM address and feeds it to the weight memory. For multi-layer SNNs or sparsity connections, the *Flit Extractor* provides the read enable (RE) signal for different layers or different links used in the weight memory. As a result, a node can have multiple AERs at the same address but for other neurons.

The LIF array's output spike is fed into the AER decoder, which extracts the address of one bit (firing neuron). This address is then serially sent to the remap Look-Up-Table (LUT) to obtain the AER value in the receiving nodes.

8.4   R-NASH Learning

As we earlier mentioned, neuromorphic systems feature asynchronous and independent execution of neurons within a neural network, therefore enabling more flexibility, as well as the ability to learn the timing information. R-NASH platform supports two learning methods: (1) off-chip learning based on a straightforward approach to take the parameters of a pre-trained ANN and map them to an equivalent-accurate SNN, and (2) online learning based on the STDP approach.

8.4.1   Off-chip Learning

For off-chip learning, we adopt the method in [11]. The feed-forward neural network is a fully connected model with a RELU (rectified linear units) activation function.

It is trained as usual using back-propagation with zero bias throughout the training. When the training is complete, we map the network's RELU weights to the IF (Integrate and Fire) network. After that, the weights are normalized and converted to SNN. Finally, the converted weights are mapped to our R-NASH model to perform inference. Note that there is no refractory and leaky used in this conversion. After being normalized, the weights are quantized into a fixed bit format and loaded to the R-NASH system via a host CPU. In particular, we use 8-bit as the de-facto format in our system. However, adopting smaller bit-width makes it possible to reduce the overall area cost because memories take up a significant portion of the system.

Since offline learning can be done with different approaches such as conversion from ANN/CNN [26, 28], learning directly with SNN [32, 33], or bio-inspired learning [16] (i.e., STDP, SDSP), different approaches can be adopted for our R-NASH system.

8.4.2   Online Learning with STDP

Despite being able to load pre-trained weights and parameters to perform only inference on our R-NASH, online learning is also supported. Here, the online unsupervised STDP with winner-take-all mechanism is adopted [10]. Once a neuron fires, it goes to the refractory mode, and an inhibitory spike is broadcasted to others to reduce the membrane potential. Figure 8.6 shows our system using STDP learning. The input spikes can be stored in SRAM and loaded to the system or generated by the host CPU. Our 3D-NoC interconnect performs the transmission of spikes. Once a neuron fires, it sends the spike to the host CPU to be counted. In the

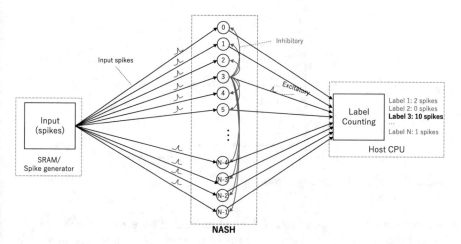

Fig. 8.6 On-chip STDP learning model

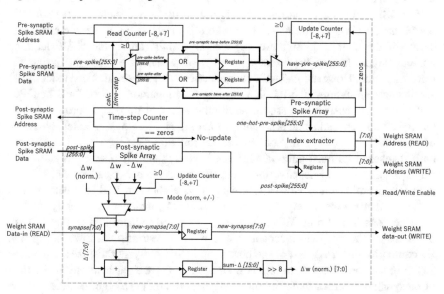

Fig. 8.7 On-chip STDP learning architecture

end, the label with the maximum number of spikes becomes the selected label. On the other hand, once a neuron fires, it sends inhibitory spikes to other neurons to reduce their membrane potential. As this STDP is unsupervised learning, the host CPU label is based on the input spike label. Figure 8.7 shows the online STDP learning block. To reduce complexity, We only adopt a simple STDP mode where the weight of synapses is adjusted to a fixed value based on the presynaptic spike's relative arrival time. If the presynaptic spike from neuron i arrives before the event of a postsynaptic spike of neuron j, the synapse weight between neuron i and j (w_{ij})

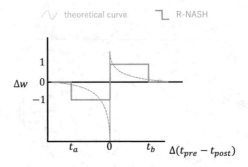

Fig. 8.8 On-chip STDP learning mechanism

is increased by a fixed Δw value. On the other hand, if the event of the presynaptic spike from neuron i arrives after the event of a postsynaptic spike of neuron j, the weight is reduced by a fixed Δw value.

R-NASH implements various reconfiguration using two methods: *(1) adaptive threshold:* once the neuron fires, its threshold is increased by a specific range, but decays if the neuron doesn't fire, *(2) weight normalization:* the STDP learning module targets having the average weights of a neuron unchanged during the learning period. However, R-NASH cannot deliver a high resolution due to hardware architecture limitations like the floating-point computation unit of CPU or GPU. For adaptive threshold, we use an additional adder in the LIF neuron to adjust the threshold. The adjustment value $(\Delta Thres.)$ in Fig. 8.5 is selected based on whether the neuron fires or not. As the neuron fires, the threshold is increased until it reaches its maximum value. Else, it decays until it reaches its minimum value. Thus, using an adaptive threshold, the neuron's firing pattern can balance with the incoming rates. Howbeit, unbalanced weights can lead to a neuron with higher weight values having a maximum firing rate, which consequently inhibits other neurons from firing. This makes the system fail to learn in a winner-take-all mechanism. The updating mechanism of the STPD learning is shown in Fig. 8.8. By comparing the timesteps of the postsynaptic and presynaptic spikes, the online STDP block can group them into two categories: (1) weight increase: if the time of presynaptic neurons is before the postsynaptic, and (2) weight decrease: if the time of presynaptic neurons is after the postsynaptic. Here, we illustrate with 16 timesteps and fixed weight change to reduce the overall complexity. Furthermore, the value can be reprogrammed via the memory interface provided by NI.

8.5 R-NASH Initial Mapping

This section presents the mapping method of the R-NASH system hardware. As we break the neuromorphic system into groups of neurons connected via a Network-on-Chip, dividing and placing are essential issues since they can heavily affect performance. For instance, placing two connected neurons far apart can lead to a critical delay path in the system. Consequently, the system needs to wait for the spike to travel a long distance before forwarding it to a new timestep. Unfortunately, this also increases the power consumption and introduces more thermal dissipation in the packet-switching network.

As R-NASH uses a genetic algorithm (GA) as the tool for generating the neuron's position and the configuration, we first cover the GA, and then we will show the GA method for R-NASH.

8.5.1 Genetic Algorithm

A Genetic Algorithm (GA) is a method for solving both constrained uncon-strained optimization problems. As we later show, the GA for initial mapping is an unconstrained one. GA relies on the natural selection process mimicking biological evolution. John Holland firstly developed a Genetic Algorithm, and his colleagues [17] is a computation method of biological evolution based on Charles Darwin's theory of natural selection. GA relies on random search and rank-based selection.

Genetic Algorithm operates based on string structures that are evolving. This string structure has a specific set of surviving rules. In each generation, the set of surviving rules is defined to remove a sub-set of members which leave a part of them in the population. GA goes through a crossover process to generate offsprings from parents to maintain the same number (or a specific number) of members. Usually, an offspring shares part of its string structure with its parents. Moreover, the mutation process help change the string structure in a stochastic manner. Genetic Algorithm usually goes through several steps: (1) Initialization; (2) Selection; (3) Crossover and (4) Mutation. The selection, crossover, and mutation operate in iterations which are called generations.

8.5.1.1 Initialization

In general, GA randomizes the initial population to have a starting one. The string structure of the GA population member is randomized under their optional constraints. To have a better initialization process, the initial population can be larger than the population in generations. Figure 8.9a illustrates the string structure for a genetic algorithm where each structure consists of six binary bits (0 or 1). In the initialization process, GA randomizes to have M members as shown in Fig. 8.9b. The size of the structure and the string format can be varied between applications and configurations.

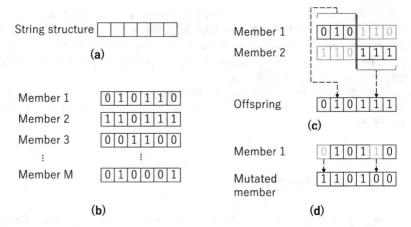

Fig. 8.9 Genetic algorithm (GA): (**a**) an example string structure; (**b**) initialization process; (**c**) cross-over and (**d**) mutation

8.5.2   Selection

A set of surviving rules defines the selection. In general, GA uses cost function or fitness function to select the survival members of the population. The fitness function (reversed value of cost function) defines the survival member by ranking from lowest to highest values.

There several ways to select surviving members. The simplest way is to rank the whole generation to select. An alternative method that can be used is tournament selection, where each tournament selects a sub-set of the population to select surviving members. After a fixed amount of tournaments, we can obtain the desired number of surviving members.

8.5.3   Crossover

Crossover in GA is based on mixing two or more parents to generate offsprings. An offspring's string structure is constructed from its parents' structure. For example, the first half of one parent and be mixed with the second half of another parent to obtain offspring. This crossover is to generate a new member that lies between the range of two members. The example of the crossover will be shown in our GA method. Figure 8.9c illustrates an example of crossover between *Member 1* and *Member 2*. Here, GA takes 50% of each parent to generates offspring.

8.5.4 Mutation

The mutation process is to alternate a member's string structure by randomizing a part of the structure. This helps spread the member to a new local range. Note that the mutation must follow the constrained set in the initialization. The example of mutation will be shown in our GA method. Figure 8.9d illustrates the mutation of a member. Here, GA randomizes the indexes, which are first and fifth, and alternate the value in the string structure (0–1 and 1–0).

8.5.4.1 Finalization

At the end of the genetic Algorithm (i.e., after G generations), the selection process is performed for only one survival member of the whole population.

8.5.5 Genetic Algorithm for Neurons Mapping on R-NASH Hardware

In addressing issues of mapping, several design factors must be carefully considered: (1) computation, (2) communication, and (3) memory. Since the computation on-chip is designed in nodes with this architecture, we can quickly realize that communication is the most critical problem. If the data is not fed fast enough to the computation unit, which is well paralleled, the system encounters a communication bottleneck. Since the multicore system's mapping issue is NP-hard (non-deterministic polynomial-time hard), R-NASH uses an optimization method using a simple Genetic Algorithm (GA) because ILP is NP-complete and the heuristic search is factorial. At the beginning of the Algorithm, it randomizes K mapping solutions. After having K mapping solutions, it enters G generations of improvement. In each generation, the GA performs the following steps as in Algorithm 6. During the G generations, tt first removes the incorrect mappings (i.e., requires more computing units than the designed node or has not mapped all computations). Then, the GA algorithm computes the cost function, which is the communication cost:

$$F_{cost} = \sum_{i=0, j=0}^{W} d_{ij} \times c_{ij} \tag{8.3}$$

where d_{ij} and c_{ij} are the distance and the connection status between neuron i and j. Since the data transfer is in a multi-cast manner at each node, c_{ij} is the connection between two PEs. F_{cost} is the communication cost for our GA. B's best communication costs are then selected out of K mapping solutions.

Algorithm 6 Genetic algorithm for neurons mapping

    // initialize phase
70 **S1:** load the system configuration
71 **S2:** randomize the K mapping solutions
    // evolve phase
72 **for** *(generation g_i in 1 to G)* **do**
73 | **S3:** remove the wrong mapping solutions
74 | **S4:** calculate cost function (communication cost) for each solution of the population
75 | **S5:** select the B best out of K solutions based on the cost function
76 | **S6:** mutate the B best solutions to have new K solutions
77 | **S6:** crossover the new K solutions to have new population
78 |_ **S7:** check if it satisfies the communication cost or does not improve over several generations

    // finalize phase
79 **S7:** calculate cost function for each solution of the population
    S8: select the $B = 1$ best out of K solutions based on the cost function

---

We can consider the communication cost as the reverse of the fitness function.

After having B best solutions, they are crossed over to obtain K solutions. Here, we keep the original B solutions and create a K-B solution using the crossover. The crossover method is shown in Fig. 8.10a. Assuming the crossover probability is 0.4 (randomly picked), the offspring take 0.4 of a parent and 0.6 of another parent to generate its mapping. Figure 8.10a shows two parents' mappings with configured neurons for [L1, L2, L3] (layer of neural net) as [100, 100, 50] and [40, 30, 80]. After the crossover process, an offspring is generated as 40% of [100, 100, 50] plus 60% of [40, 30, 80] which is [76, 72, 62].

After the crossover generates K solutions, the GA mutates the K solutions to generate mutated configurations (under a certain probability). Figure 8.10b shows how we mutate a configuration. Since the number of neurons mapped in each layer and in each PE is constant, we must maintain it. In Fig. 8.10b we describe a case where the L3 of the node (0,1,0) is randomly mutated. To maintain the number of neurons mapped, we randomize the L2 of the node (0,2,3) for processing. We find that the minimum of two configurations is 30, which means we can reduce both configurations by 30. Meanwhile, by reducing L3 of (0,1,0) by 30, we increase L2 by 30 and L3 of (0,2,3) by 30. Note that the reduction value can be randomized between 0 and 30. However, our experiment works best with the minimum one.

After mutation, we re-update the configuration to match the number of unused neurons. Then, we check whether we satisfy the communication cost in the specification. If the communication cost is good enough, we can end the mapping. The GA is also completed after G generations.

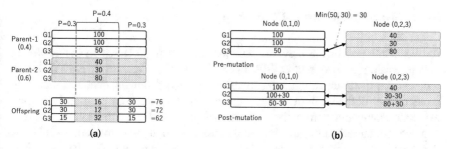

Fig. 8.10 Crossover and mutation method for GA mapping. (**a**) Crossover. (**b**) Mutation

8.6   R-NASH Run-Time Maintenance

As previously discussed, robustness is the primary target of R-NASH neuromorphic hardware. Therefore, R-NASH also provides a comprehensive set of fault-tolerance features. In general, there are three significant parts of the R-NASH system that need to be protected: (1) data integrity, (2) interconnect, and (3) computing engine. First, we discuss how R-NASH protects against data corruption. Then we describe the fault tolerance feature for the communication in R-NASH. Finally, uncorrectable faults are discussed and recovered from.

8.6.1   Data Integrity Protection

One of the most basic protections in any highly reliable system is to be resilient against data corruption. Here, we classify the data into two types: (1) one-time load data and (2) transferring data. While one-time load data such as weights and thresholds are essential, their accuracy can be maintained by storing them in on-chip SRAM or registers. Here, an error correction code can be used to protect these types of data. However, from our investigation, we discovered that soft errors have little impact on the overall accuracy. Therefore, these types of data can be left unprotected and can periodically be written to ensure correctness. The other type of data is the one that is transferred among the system. In particular, spikes and synchronizations are significant types of flits. The initialization of the system is also essential. Synchronization flits are critical since they ensure the operability of the system. To protect this type of data, we embed on our 3D-NoC, two sets of SECDED (22,16) [18]. As a result, in transferring 32-bit of data, the 3D-NoC needs to transfer 44 bits, as this allows our R-NASH to be resilient against one fault per set and be alert against two faults per set. In other words, our 3D NoC allows 2-bit correction and 4-bit detection at its best. By protecting the data, we can ensure the system works with a good level of confidence.

8.6.2 Communication Protection

As we mentioned in the previous section, the data is protected in our R-NASH system, thanks to two sets of SECDED (22,16) [18]. Moreover, R-NASH also protects the communication infrastructure with the following features. First, by protecting the defective buffer, a technique named Random Access Buffer [1] is used to isolate a faulty buffer from the read and write process. Furthermore, if the crossbar is defective, there is a backup link in the crossbar to allow communication [1]. If the input port, output port, or the whole router is defective, a fault-tolerant routing algorithm can recover the system [1].

8.6.3 Fault-Tolerant Neurons Mapping Scheme

Another issue in large-scale SNN architecture is that modules can develop uncorrectable faults during runtime. This means that the module is corrupted and cannot be used to obtain even graceful accuracy degradation. Therefore, We present a method to remap the neuromorphic system under such faulty circumstances. To protect the faulty neural computing unit against defects, we use two strategies: (1) a node of neuron has some spare neurons (and their weight SRAM), and (2) there is a spare node in the system. Once a neuron/node fails, R-NASH can remap the neuron/node to the spare one and keep its operation. Figure 8.11a shows an example of a layer in this configuration. Here, each node has a different number of spare neurons. Once there are failed ones, R-NASH maps them to spare neurons. Figure 8.11b shows the 25 faulty neurons of node (0,0,0) remapped into node (0,3,2). The number of spare neurons in a node (0,3,2) is reduced from 34 to 9 neurons. Since the mapping method already exists, we can use the usual SNN mapping method for replacing neurons. However, the remapping of SNN can provide an alternative approach since there are new factors in faulty situations. For instance, we might want to reduce the downtime (repairing time) or limit memory transfer within the system. On the other hand, disconnected regions due to faulty network sections can be problematic for mapping. Due to these reasons, we also consider the conventional method: *(1) Greedy Search:* all faulty nodes run once to find the replacement, and *(2) Genetic Algorithm* by adjusting the existing Algorithm for mapping to obtain a more suitable solution. In the scope of this chapter, we present the adaptation of the Genetic Algorithm for remapping. Note that three approaches are also implemented and compared in the evaluation section. Besides the communication cost in Eq. 8.3, we introduce the migration cost to reduce the repairing time of the system as follows:

$$M_{cost} = \sum_{i=0, j=0}^{W} d_{ij} \times m_{ij} \tag{8.4}$$

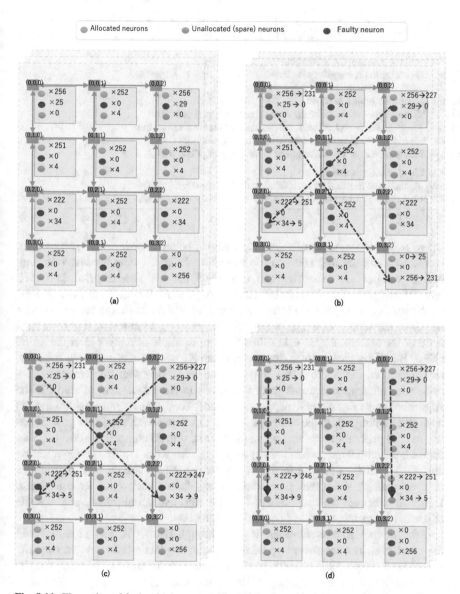

Fig. 8.11 Illustration of fault-tolerance remapping of the Genetic Algorithm. (**a**) Unmapped and free neurons per node. (**b**) A randomized mapping solution. (**c**) Mutating by finding a shorter distance for a flow; (**d**) Mutate by swapping destination of a flow

---

Algorithm 7 Genetic algorithm for remapping SNN

      // initialize phase
80 **S1:** build the unmapped and free neurons per node
81 **S2:** randomize the K mapping solutions
      // evolve phase
82 **for** *(generation g_i in 1 to G)* **do**
83 | **S3:** calculate cost function for each solution of the population
84 | **S4:** select the B best out of K solutions based on the cost function
85 | **S5:** mutate the B best solutions to have new K solutions
86 | **S6:** crossover the new K solutions to have new population

      // finalize phase
87 **S7:** calculate cost function for each solution of the population
88 **S8:** select the $B = 1$ best out of K solutions based on the cost function

---

where m_{ij} is the number of migrating neurons between node i and j. The main reason to adopt the migration cost is to reduce the repair time because reloading neuron weights are expensive and can affect the real-timeliness of R-NASH. Algorithm 7 shows the Remapping Genetic Algorithm. It consists of three phases: (1) initialize, (2) evolve, and (3) finalize. The initialize phase starts with the first step **S1** where the number of unmapped and free neurons are counted and sent from each node of the system. Figure 8.11a illustrates an example of a layer after the initial phase. Based on these values, the second step **S2** generates K mapping solutions randomly (i.e., Fig. 8.11b). This step randomizes a node with free neurons and a node with unmapped neurons from the values in **S1**. At the end of step *S2*, the Algorithm generates K legal mapping solutions. They are not optimal solutions and need to be optimized. In the *evolve* phase, the GA method iterates for G generations where each generation repeats four steps. At first, step **S3** compute the cost function for each solution. Here, we can adopt only M_{cost} from Eq. 8.4. The communication cost F_{cost} is also calculated for the selection step $S4$. In $S4$, the best B solutions in K are ranked, and if they have similar M_{cost} values, their F_{cost} are considered. Doing so preserves the simplicity of a single objective optimization for GA while still considering migration and communication costs.

After getting B best solutions, it goes through two steps: **S5**—crossover and **S6**—mutation. The crossover step **S5** is performed by mixing two random mapping solutions. It takes 50% of each parent to generate offspring. By doing so, the offspring can inherit the mappings of its two parents.

There are two types of mutations in the mutation step **S6**. First, it finds an immediate random node between two random nodes having a mapping flow. Here, we ensure the immediate node having free neurons is closer to the source node than the flow destination. For instance, Fig. 8.11b shows the case where the source node $(0, 0, 0)$ has 25 unmapped neurons and all are mapped to $(0, 3, 2)$—the destination node. Then, it finds the immediate node $(0, 2, 2)$ with two conditions: (1) there are free neurons in the immediate node $(0, 2, 2)$ and (2) the distance from the source node $(0, 0, 0)$ to the immediate node $(0, 2, 2)$ is smaller than the original flow $((0, 0, 0)$ to $(0, 3, 2))$. The neurons are then remapped to the immediate node instead of the destination. The result can be seen in Fig. 8.11c.

The second mutation is to swap the mapping to have a closer migrating distance (smaller M_{cost}). If two flows can have a smaller migrating distance by swapping the destination, the Algorithm performs the swap. For instance, Fig. 8.11c shows unmapped neurons in node $(0, 0, 0)$ being mapped to $(0, 2, 2)$, and unmapped neurons in node $(0, 2, 0)$ being mapped to $(0, 0, 2)$. Here, the migrating distances are four for both flows. However, by switching the destination, we obtain a migrating distance of two for both flows, as shown in Fig. 8.11d.

After G generations, the Algorithm finalizes by selecting only the best solution (step **S7** and **S8**). This solution is used to perform the mapping method. Since the GA might take a long time to complete, we can also allow early termination of the mapping and use the best-found solution.

In summary, this GA methodology provides an extension for the optimization problem of remapping faulty neurons. While the mapping algorithm only focuses on the communication cost, GA allows designers to take the migration cost function for the optimization.

8.7 R-NASH Evaluation Results

In this section, we evaluate our R-NASH platform. First, the initial mapping issue is addressed to show the efficiency of the GA (Genetic Algorithm) model. Here, we map multiple-layer feed-forward networks to different 3D NoC sizes from $4 \times 4 \times 4$ to $10 \times 10 \times 10$. To understand the effect of having different node sizes, we map the same system into different node sizes (256, 128, 64, and 32 neurons per node) and topologies. We also investigate the difference between 3D and 2D topology to illustrate the benefits of 3D structure. To improve the robustness of the neuromorphic hardware, the fault-tolerant mapping is presented compared to conventional works like greedy search. Third, we present the hardware complexity for our system with NANGATE 45 nm and FreePDK45 TSV library. We then present both offline and online training for the MNIST dataset in our R-NASH neuromorphic hardware. To validate the operation of R-NASH, the tasks were performed with both offline and online learning. For offline learning, offline trained weights of a multiple-layer feed-forward neural network were converted and used for classification on R-NASH. Thanks to the inhibitory connections, the task with online learning was performed using the on-chip STDP with a winner-take-all mechanism. Finally, we discuss the pros and cons of our approach.

8.7.1   Initial Mapping Evaluation

8.7.1.1   Mapping over Different 3D-NoC Sizes

To understand the GA method's efficiency for initial mapping, we first compare it with three linear mapping solutions. Then, we adopt the linear mapping method from SpiNNaker [19] and implement it for the 3D topology to get Linear X, Linear XY, and Linear XYZ, which represent the priority direction of the linear mapping. The mapping configuration can be found in Table 8.1. Note that there are fixed spare neurons (20% per node) and an extra node (the highest index node) that are not used for mapping and can be used to tolerate faults later in Sect. 8.7.2. Here, we run GA with population $K = 100$, the best $B = 5$, and the mutation rate of 0.5. Figure 8.12 shows the result of GA in comparison to the linear mapping methods. With a small network size of $4 \times 4 \times 4$, after nearly 60 generations, the GA saturates at a point, and the lowest communication cost stays unchanged over the rest of the generations. The final communication cost is lower than both manual mapping solutions, and the overall cost is $0.4\times$ the manual mapping. With bigger NoCs, it is easy to understand that it needs more generations to be lower than the linear mapping. While $4 \times 4 \times 4$ takes around 60 generation to converge, $6 \times 6 \times 6$, $8 \times 8 \times 8$, and $10 \times 10 \times 10$ need around 120, 180, and 320 generations respectively to be stable. We can observe that linear mapping methods have significantly higher communication costs in all tested cases than the genetic algorithm ones.

8.7.1.2   Mapping over Different Node Sizes

Figure 8.13 illustrates the different node sizes from 32 to 256 when mapping for the same system. To maintain the same system size (the same number of neurons), we vary the 3D NoC size from $4 \times 4 \times 4$ to $8 \times 8 \times 8$. As we can observe in Fig. 8.13, the smaller NoC benefits the smaller distances between nodes, which can reduce the cost of communication. The smallest network size provides the lowest

Table 8.1 Configuration for the mapping evaluation[a]

Parameter	Value
# neurons per node (E)	256
# nodes (N)	$4 \times 4 \times 4$ to $10 \times 10 \times 10$
# spare neurons (R)	$0.2\times X$
# spare node	1
# faults (k)	$0.05\times X$, $0.10\times X$, $0.15\times X$, and $0.20\times X$
SNN # layers	4
SNN configuration[a]	784:0.5*(W-10): 0.5*(W-10): 10

X: number of neurons in R-NASH

[a] MLP model. For example, the SNN configuration for $E = 64$ and $4 \times 4 \times 4$ is 784:1633:1633:10

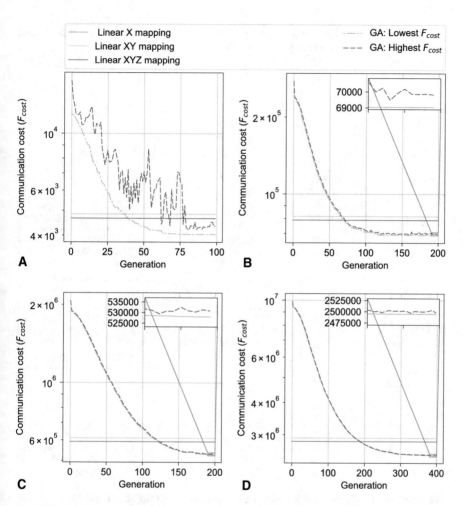

Fig. 8.12 Genetic Algorithm Result for initial mapping. (**a**) 4 × 4 × 4 NoC-based, 256 neurons/node. (**b**) 6 × 6 × 6 NoC-based, 256 neurons/node. (**c**) 8 × 8 × 8 NoC-based, 256 neurons/node. (**d**) 10 × 10 × 10 NoC-based, 256 neurons/node

communication cost (4039). By increasing the network's size and reducing the node's size, the communication cost keeps increasing. With 128, 64, and 32 neurons per node, the communication costs are 21,768, 107,838 and 529,440, respectively. We do not need to send multiple unicast flits for spikes by placing neurons in the same layer into a node. Instead, sending a single flit and distributing it to all nodes can significantly reduce the traffic. However, we would like to note that scaling the number of neurons per node is not unlimited due to the limitation on crossbars and bottleneck on on-chip communication. Moreover, having a large size node also leads to some disadvantages: (1) lower operating frequency due to a long critical path; (2) difficulty to place and route due to complex structure and macro SRAM

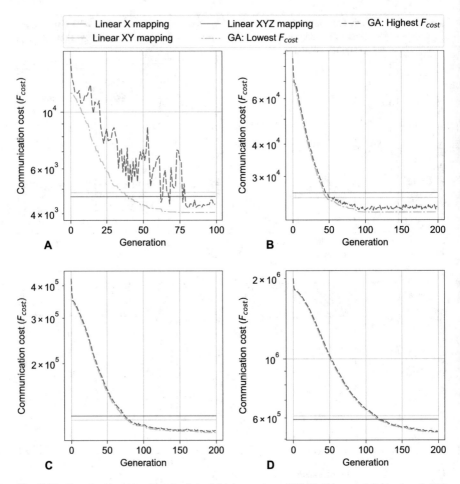

Fig. 8.13 Genetic Algorithm Result of the initial mapping of 3D NoC-based. (**a**) $4 \times 4 \times 4$, 256 neurons/node. (**b**) $4 \times 4 \times 8$, 128 neurons/node. (**c**) $4 \times 8 \times 8$, 64 neurons/node. (**d**) $4 \times 8 \times 8$, 32 neurons/node

and (3) long-distance between nodes could also reduce the performance of the NoC. Typically, the neuromorphic cluster varies between 256 and 1024 neurons per node.

8.7.1.3   Comparison Between 3D and 2D in Initial Mapping

As we mentioned, the 3D structure brings benefits. To illustrate them, we compare the communication cost between 3D and 2D under the same linear mapping and GA in Fig. 8.14. We keep the same node size as 256 neurons per node for a fair comparison and the change between the NoC sizes. We compare 3D and 2D NoC with the same number of nodes (64, 128, 256, and 512). As can be observed in

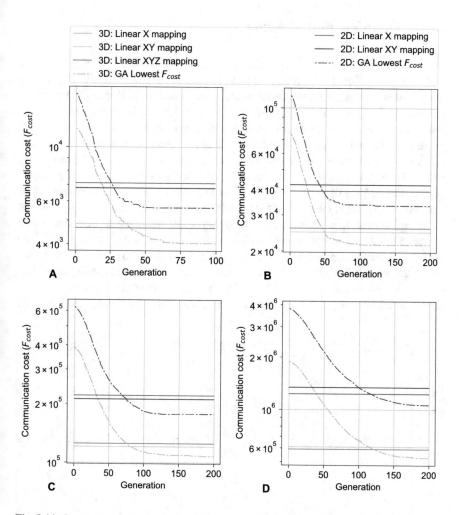

Fig. 8.14 Comparison between 3D and 2D mapping. (**a**) 64 nodes (4 × 4 × 4 and 8 × 8) NoC-based, 256 neurons/node. (**b**) 128 nodes (4 × 4 × 8 and 8 × 16) NoC-based, 256 neurons/node. (**c**) 256 nodes (4 × 8 × 8 and 16 × 16) NoC-based, 256 neurons/node. (**d**) 512 nodes (8 × 8 × 8 and 16 × 32) NoC-based, 256 neurons/node

Fig. 8.14, mapping on a 3D structure leads to a significantly smaller communication cost. In all test cases (64, 128, 256, and 512 nodes), the performance of GA on 3D is 1.4–2.0× smaller than the 2D ones. Even with linear mappings (X, XY, or XYZ), 3D still dominates the 2D. This is due to the nature of 3D bringing much shorter traversal paths between regions of the chip. As a result, spikes can travel much faster on 3D, and this could be translated into better performance and lower power in 3D.

8.7.2   Fault-Tolerant Mapping

In this section, the *Genetic Algorithm* for remapping is evaluated and compared with 1-hop and N-hop *Greedy Search* (GS) to understand its efficiency. The *Greedy Search* runs each node once and looks for a spare node within one (1) hop range or in the entire system (N-hop) with the shortest distance. The configuration of the evaluation is shown in Table 8.1. In this evaluation, we measure three significant parameters: *(1) mapping rate:* the ability to map the faulty neurons to the spare ones; *(2) average spike transmission cost (F_{cost}):* the average distance of all connections and *(3) Migration cost M_{cost}: the amount of read/write neurons needed to adapt the system.* Figure 8.15 illustrates the results for the system for 3D-NoC configurations (see Table 8.1). As shown in Fig. 8.15, our GA method can map all faulty neurons to the spare ones regardless of the size or topology. We have to note that 1-hop *Greedy Search* can only map around 60% (around 80% with the worst cases) of the faulty neurons. This is because 1-hop *Greedy Search* only runs once and looks for one mapping solution of its neighbor to fail to map easily. Meanwhile, the N-hop *Greedy Search* and the Genetic Algorithm can map all neurons. The average F_{cost} (communication cost) also varies between different approaches. Since the 1-hop GS

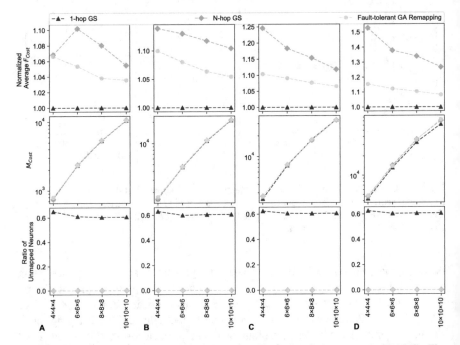

Fig. 8.15 Output mapping for the migrated neurons with random fault patterns in 3D-NoCs. The system has 256 neurons per node; 20% of the neurons are spares with 1 redundant node without any allocated neuron at 0% fault rate. (**a**) 5% defect rate. (**b**) 10% defect rate. (**c**) 15% defect rate. (**d**) 20% defect rate

mostly fails to map the neurons, the average communication distance per neuron is unchanged. For other methods, the average F_{cost} fluctuates between different sizes. However, as we can observe in Fig. 8.15, they are reduced when we increase the size of the NoC. This is because when we increase the size of the NoC, the impact of moving neurons is reduced. The effects are also more minor, with smaller fault rates (k values). We can even notice the communication cost is maintained with remapping. However, a slight reduction can be observed with the migration-based Algorithm. Also, GA seems to have a better average F_{cost} since it reduces that value as the second factor. In conclusion, we have shown the efficiency of adapting the GA (Genetic Algorithm) for solving the remapping problem. Although the GA in some cases still has a higher F_{cost} than others, it has shown efficiency on both migration cost and communication cost. Moreover, it offers efficiency even with high defect rates where the communication cost is much lower.

8.7.3 Hardware Complexity

Table 8.2 shows the hardware complexity of the architecture. The NI, which supports the mapping method, is integrated with the neuron cluster and the 3D-NoC router. As shown in Table 8.2, the additional LUTs for AER and Address take up 23.35% and 28.83% of the Network Interface area, respectively. The overhead of these two LUTs is relatively small. On the other hand, the NI, which supports migration techniques, only occupies 25.95% of the tile area without the SRAM (neuron cluster + network interface). Figure 8.16 illustrates our sample layout for a 2 × 2 NoC-based SNN layer with migration support. The cluster's configuration is 256 spike inputs in AER format, 8-bit synapse weight, 32 physical neurons, 32 synapse crossbars for each cluster. Here, each crossbar is implemented with a 256-bank 8-bit dual-port SRAM using OpenRAM. We only integrate 32 neurons per node to have a reasonable Place&Route time and a visual layout. To support 3D-NoC inter-layer interconnect, we use TSV from FreePDK3D45 with the size of

Table 8.2 Hardware complexity of the R-NASH node

Module	Area (μm^2)	Power (mW)	Max Freq. (MHz)
Network interface			
AER LUT	16,747	–	–
Address LUT	20,768	–	–
Total	72,032	30.4043	699.30
Neuron cluster	205,608	81.682	751.87
	64KB SRAM	–	–
3D-NoC router [7]	41,739	14.6128	537.63
Vertical TSVs (up and down)	2901.1136	–	–

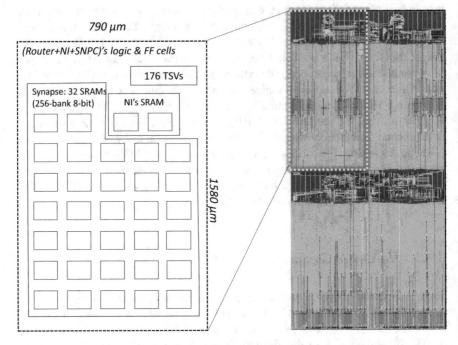

Fig. 8.16 Layout of a 2 × 2 3D-NoC-based R-NASH layer. A tile's size is 790 μm × 1580 μm

4.06 μm^2 × 4.06 μm^2 and the Keep-out-Zone is 15 μm^2 × 15 μm^2 for each TSV. As can be observed in the layer's layout, 80% of the area is for placing macro SRAM. Since the design of the LIF neuron is lightweight, the most complicated part is the crossbar.

However, the NI requires two dedicated SRAMs to convert the AER from local value to a global one and a destination lookup table. We can further optimize the design's footprint by reducing the bit-width of a synapse or using an alternative memory approach (eDRAM, STT-RAM, or memristor). Moreover, we add more stacking layers dedicated to memory, which allows us to have a smaller footprint.

8.7.4 System Validation

This section presents the result of online and offline training for our R-NASH system. As the 3D-NoC aims to model a complex neuromorphic system, the conventional MNIST classification neural networks are too small to map. Therefore, we have scaled up the feed-forward neural network's hidden layer to map it into our R-NASH in the validation section. We first show the offline conversion from a feed-forward ANN to our SNN. Next, weight, parameters, and configurations are loaded into the system using a memory interface. Then, the online STDP method is presented. The initial weights, randomized and normalized, are loaded in our R-NASH system as R-NASH does not support random number generation.

8.7.4.1 Offline Feed-Forward Network

For the offline training, we use the feed-forward network 784:1024:1024:10 and 784:1024:1024:1024:10 for the MNIST dataset. We fixed the hidden layer to 1024 to fit the 10-bit SRAM model for the hardware design. Here, we use a 3D-NoC of $4 \times 4 \times 4$ with 64 neurons/node from the output of the initial mapping in Fig. 8.13. Since 2058 and 3082 neurons are used in the two networks, we reserve the remaining ones as the spare neurons for tolerating potential defects. There is one extra node at (3,3,3) and sparse neurons of around 31/32 for the first network and 2/3 for the second network in all active nodes. Although scaling the 3D-NoC and the number of neurons per node can support bigger network size, we only adopt the above feed-forward size to avoid large SRAM models. Using sparse synapses could reduce the SRAM size; however, we only target to validate R-NASH's operation.

Figure 8.17 show the accuracy results of 784:1024:1024:10 on the R-NASH system in comparison with the software version. The R-NASH system uses an 8-bit signed weight representation which gives similar results to the converted version in Matlab. The total number of time steps is 350 (1 ms per time step in the simulation). Here, we also evaluate the fixed point SNN in software where we clip the least significant bit in representation. We also consider our R-NASH, where

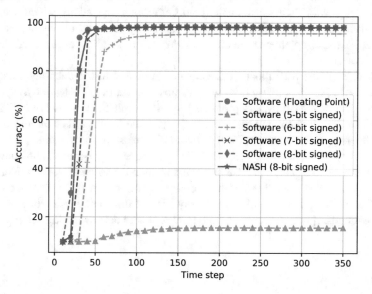

Fig. 8.17 Accuracy result of offline training for MNIST dataset with the network model 784:1024:10

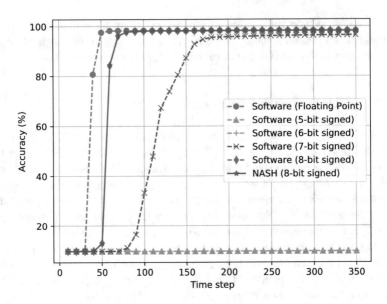

Fig. 8.18 Accuracy result of offline training for MNIST dataset with the network model 784:1024:1024:10

8-bit signed fixed-point values are converted to an integer value to enable hardware implementation. At first, we can easily see the drop in accuracy when comparing the floating-point SNN and the fixed point ones. The reductions are significant when the number of representing bit is less than 5. The main reason is that the more extensive and deeper network will accumulate the differences in the values, which results in more inaccurate results. Nevertheless, we can easily see that an 8-bit signed fixed point is best for implementation and provides nearly identical accuracy at the end with a slower response time.

Figure 8.18 and 8.19 illustrate the case of three and four hidden layer network (784:1024:1024: 1024:10). Here, we can observe a similar behavior as the first network. The R-NASH system provides a similar result as the software in floating-point. The 7-bit fixed point version now can have the final result of inference close to the floating-point; however, it needs over 100-time steps to converge.

In summary, we illustrate that R-NASH can inference pre-trained and converted networks without issues, and the result is identical to the software version. Note that R-NASH saturates around the 55th time step in both cases. If the system cuts the operation at this point, it could save nearly 85% of the computation time by using clock gating [22] where energy could be saved at zero data switching activity.

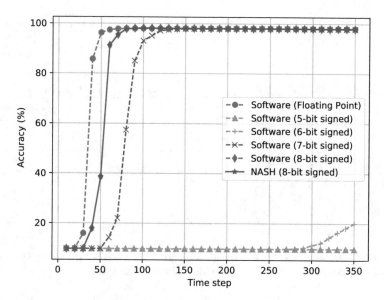

Fig. 8.19 Accuracy result of offline training for MNIST dataset with the network model 784:1024:1024:1024:10

8.7.5 Unsupervised STDP

In this section, we evaluate the online STDP method for the same MNIST benchmark. Here we adopt the network in [10] with the recurrent version that could be found in the work of [16]. Furthermore, we simplify the architecture to be identical to the hardware implementation. The network size of [10] is 784:N:N while our network is 784:N. Since the number of neurons is not significant enough to scale to a 3D-NoC, we only use $1 \times 4 \times 4$ 3D-NoC and 64 neurons/node. There are two versions with N = 100 and N = 400, mapped into 2 and 4 nodes. Since there is no sparse connection, the communication cost stays unchanged with any neuron placements.

For testing purposes, we adopt the BindsNet [16] platform to build the RTL-like version of the LIF neuron to train and test on our PC. After completing the testing and debugging phase, the Verilog model train is performed and compared with the golden reference software. In the software model of SNN [10], the authors used the adaptive weight change ($\Delta w = w \times$ learning_rate); however, it is not suitable for our hardware STDP due to two reasons: (1) the resolution of the weight (8-bit) is too small to use the same principle and (2) the architecture for the multiplication is too complicated. Therefore, we use the fixed weight change here. We also evaluate the method for more understanding. Table 8.3 shows the accuracy of the software version of STDP learning and our hardware STDP SNN. Comparing the R-NASH model and the software [10], we could observe a drop in accuracy by using our

Table 8.3 Accuracy result of STDP learning for SNNs

N	Floating point software	R-NASH
100	79.44%	71.32%
400	88.87%	84.05%

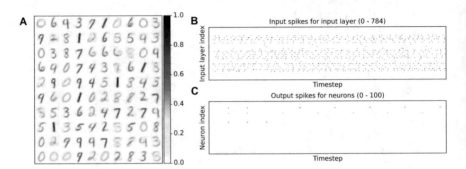

Fig. 8.20 Illustration of the STDP learning model. (**a**) the final weights. (**b**) Illustration of input spikes for the first test image (number 8). (**c**) Illustration of output spikes

RTL model. This is due to the much simpler hardware model and lower resolution (fixed 8-bit for weight, 16-bit for membrane potential, 16-bit for normalizing).

Figure 8.20 illustrates the weight with $N = 100$ and the input (output) spikes extracted from our R-NASH software model. The weights have been adapted into the MNIST. However, there are some drawbacks due to the hardware model's simplicity. For instance, there is some weight with a similar distribution. As a result, these neurons fire simultaneously and continue to fire during the following time steps. Figure 8.20 illustrates that three neurons continue to fire during the 350 simulated timesteps. Moreover, the weights keep changing during the training time due to the pure STDP without intervention. We can certainly observe that some weights are mixed of two numbers (i.e., 8 and 5, 1 and 7).

8.8 Chapter Summary

This chapter presents R-NASH, a robust 3D neuromorphic system that supports high parallelism, thanks to the used 3D-NoCs. Besides the benefit of a smaller footprint brought by 3D-ICs, the R-NASH system is designed with highly reliable features that support data protection and fault recovery at the system level. In this work, we presented the functionality of the 3D neuromorphic system by performing the MNIST dataset classification. Moreover, the platform for R-NASH is also presented with the mapping method and fault-tolerance feature. The mapping method shows that it can easily outperform manual mappings by optimizing the number of hops needed to travel among the SNN. On the other hand, we also illustrate a genetic algorithm for fault recovery in SNN. Further optimization, such as bit-width reduction and low-power optimization, is needed and investigated in our future work.

References

1. Ahmed AB, Abdallah AB (2014) Graceful deadlock-free fault-tolerant routing algorithm for 3D Network-on-Chip architectures. J Parallel Distrib Comput 74(4):2229–2240
2. Akopyan F et al (2015) TrueNorth: Design and tool flow of a 65 mW 1 million neuron programmable neurosynaptic chip. IEEE Trans Comput Aided Des Integr Circuits Syst 34(10):1537–1557
3. Banerjee K et al (2001) 3-D ICs: A novel chip design for improving deep-submicrometer interconnect performance and systems-on-chip integration. Proc IEEE 89(5):602–633
4. Ben Abdallah A, Dang KN (2021) Toward robust cognitive 3d brain-inspired cross-paradigm system. Frontiers Neurosci 15:795
5. Ben Ahmed A, Ben Abdallah A (2013) Architecture and design of high-throughput, low-latency, and fault-tolerant routing algorithm for 3D-network-on-chip (3D-NoC). J Supercomput 66(3):1507–1532
6. Benjamin BV, et al (2014) Neurogrid: A mixed-analog-digital multichip system for large-scale neural simulations. Proc IEEE 102(5):699–716
7. Dang KN, Ahmed AB, Okuyama Y, Abdallah AB (2020) Scalable design methodology and online algorithm for TSV-cluster defects recovery in highly reliable 3d-NoC systems. IEEE Trans Emerg Top Comput 8(3):577–590
8. Dang KN, Ben Abdallah A (2019) An efficient software-hardware design framework for spiking neural network systems. In: 2019 International conference on internet of things, embedded systems and communications (IINTEC), pp 155–162
9. Davies M et al (2018) Loihi: A neuromorphic manycore processor with on-chip learning. IEEE Micro 38(1):82–99
10. Diehl PU, Cook M (2015) Unsupervised learning of digit recognition using spike-timing-dependent plasticity. Front Comput Neurosci 9:99
11. Diehl PU et al (2015) Fast-classifying, high-accuracy spiking deep networks through weight and threshold balancing. In: 2015 International joint conference on neural networks (IJCNN), July 2015, pp 1–8
12. Eliasmith C, Stewart TC, Choo X, Bekolay T, DeWolf T, Tang Y, Rasmussen D (2012) A large-scale model of the functioning brain. Science 338(6111):1202–1205
13. Furber S (2016) Large-scale neuromorphic computing systems. J Neural Eng 13(5):051001
14. SB Furber et al (2014) The SpiNNaker project. Proc IEEE 102(5):652–665
15. JH Goldwyn, Imennov NS, Famulare M, Shea-Brown E (2011) Stochastic differential equation models for ion channel noise in Hodgkin-Huxley neurons. Phys Rev E 83:4190–4208
16. Hazan H et al (2018) BindsNET: A machine learning-oriented spiking neural networks library in Python. Front Neuroinform 12:89
17. Holland JH et al (1992) Adaptation in natural and artificial systems: an introductory analysis with applications to biology, control, and artificial intelligence. MIT Press
18. Hsiao MY (1970) A class of optimal minimum odd-weight-column SEC-DED codes. IBM J Res Devel 14(4):395–401
19. Jin X (2010) Parallel simulation of neural networks on spinnaker universal neuromorphic hardware. The University of Manchester (United Kingdom)
20. Lee HG, Chang N, Ogras UY, Marculescu R (2008) On-chip communication architecture exploration: A quantitative evaluation of point-to-point, bus, and network-on-chip approaches. ACM Trans Des Autom Electron Syst (TODAES) 12(3):1–20

21. Levin JA, Rangan V, Malone EC (2014)   Efficient hardware implementation of spiking networks. Patent No. US 2014/0351190 A1, Filed May 1, 2014, Pub. Date Nov. 27, 2014
22. Mahmoodi H et al (2008) Ultra low-power clocking scheme using energy recovery and clock gating. IEEE Trans Very Large Scale Integr (VLSI) Syst 17(1):33–44
23. Ogbodo M, Vu T, Dang K, Ben Abdallah A (2020) Light-weight spiking neuron processing core for large-scale 3D-NoC based spiking neural network processing systems. In: 2020 IEEE international conference on big data and smart computing (BigComp), pp 133–139
24. Panth SA, Samadi K, Du Y, Lim SK (2014) Design and cad methodologies for low power gate-level monolithic 3d ICS.   In: Proceedings of the 2014 international symposium on low power electronics and design, pp 171–176
25. Purves D, Augustine G, Fitzpatrick D, Hall W, LaMantia AS, McNamara J (2008) Neuroscience. Sinauer Associates
26. Rueckauer B et al (2017) Conversion of continuous-valued deep networks to efficient event-driven networks for image classification. Frontiers Neurosci 11:682
27. J Schemmel et al (2010) A wafer-scale neuromorphic hardware system for large-scale neural modeling. In: Proceedings of 2010 IEEE international symposium on circuits and systems, May 2010, pp 1947–1950
28. Sengupta A, Ye Y, Wang R, Liu C, Roy K (2019) Going deeper in spiking neural networks: Vgg and residual architectures. Frontiers Neurosci 13:95
29. Stimberg M, Brette R, Goodman DF (2019) Brian 2, an intuitive and efficient neural simulator. eLife 8:e47314
30. Vu TH, Okuyama Y, Ben Abdallah A (2019) Comprehensive analytic performance assessment and K-means based multicast routing algorithm and architecture for 3D-NoC of spiking neurons. J Emerg Technol Comput Syst 15(4):34:1–34:28
31. Waldrop MM (2016) More than moore. Nature 530(7589):144–148
32. Wu Y, Deng L, Li G, Zhu J, Shi L (2018) Spatio-temporal backpropagation for training high-performance spiking neural networks. Frontiers Neurosci 12:331
33. Yin S, Venkataramanaiah SK, Chen GK, Krishnamurthy R, Cao Y, Chakrabarti C, and Seo JS (2017) Algorithm and hardware design of discrete-time spiking neural networks based on back propagation with binary activations. In: 2017 IEEE biomedical circuits and systems conference (BioCAS). IEEE, pp. 1–5

Chapter 9
Survey of Neuromorphic Systems

Abstract Neuromorphic computing systems have the potential to realize brain-like learning and adaptation ability. This chapter presents a comprehensive survey of research works in neuromorphic computing systems. First, the chapter gives the motivations for neuromorphic computing, then describes significant research works in the field. These works are categorized as software emulation, digital hardware, and analog and mixed-signal hardware approaches. This chapter aims to provide an exhaustive review of the research conducted in neuromorphic computing and illuminate the gaps in the field where new research is needed.

9.1 Introduction

During the 1980s, Carver Mead and his team at Caltech pioneered the idea of bio-inspired microelectronics [16]. Mead's approach to neuromorphic engineering is based on the physics of transistors mimicking the operation of neurons. Since then, the architectures of neuromorphic systems have been improved by researchers to obtain better scale, learning, and efficiency.

Many large-scale neuromorphic systems have been developed in recent years due to the fact that many transistors can be integrated into a single chip. By doubling the density of a single chip every two years, the available resource for designers to develop large-scale neuromorphic system has become feasible. From the design perspective, there are three design approaches for neuromorphic systems: (1) software emulation, (2) digital hardware, and (3) analog and mix-signal hardware.

In the software emulation approach, the SpiNNaker system designed by the University of Manchester is the most notable. It is structured as a high performance computing system with many ARM cores that simulate models of neurons and synapses. A SpiNNaker chip which consists of 18 cores communicate via on-chip network that extends off-chip via wires.

Recent chips that adopt the digital hardware approach include IBM TrueNorth and Intel Loihi. TrueNorth is a multi-core Network-on-Chip-based neuromorphic

© The Author(s), under exclusive license to Springer Nature Switzerland AG 2022   217
A. Ben Abdallah, K. N. Dang, *Neuromorphic Computing Principles and Organization*, https://doi.org/10.1007/978-3-030-92525-3_9

chip. Each core is a 256 × 256 crossbar connecting the incoming spike event to the post-synaptic neurons. It deploys a Leaky-Integrated-and-Fire neuron model with a fixed weight for each neuron. Training for TrueNorth is done off-chip as the system only performance inference. The Intel Loihi consists of an asynchronous spiking neural network (SNN), which enables it to sends spikes along active synapses. Its programmable neurons allow different models to be implemented. The bit-width of synapses is also adaptive between 1 to 9 bit.

The Analog and mix-signal hardware approach inherit the idea of Prof. Carver Mead and his team at Caltech. The BrainScaleS system uses the wafer-scale analog circuit that performs 10,000 times faster than the biological real-time. The Stanford University Neurogrid with the real-time sub-threshold analog neural circuit. It's on-chip network is based on tree topology, and delivers multicasting performance.

All of these approaches represent different trade-offs between a set of desirable objectives. Factors like energy efficiency, integration density, flexibility, analog versus digital neuron, hardware versus software, all play their role in creating balance in the above systems. In the following section, we will cover these systems.

Table 9.1 shows the comparison between notable works of large-scale neuromorphic computing systems. SpiNNaker can simulate up to one million neurons. However, because it is based on high-performance computing approach, it is difficult to scale up further. On the other hand, chip-based solutions, such as TrueNorth, Neurogrid, or Loihi can be expanded with a multi-chip solution as 2D-NoC can be naturally scaled. For example, both 16 chips versions of Neurogrid and TrueNorth can reach one million neurons without consuming a significant amount of power.

Although hardware-accelerated systems prove to be energy efficient, SpiNNaker has the flexibility on the neuron operation as it can be further updated. SpiNNaker also supports better memory and plasticity model. Loihi, has a programmable neuron operation that allows the designers to change the neuron's function; however, it is easy to see that Loihi is the most complicated design as it uses 2.07 billion transistors for 131 thousand neurons. Another observation for the state-of-the-art design is the domination of digital neuromorphic chips. Although analog neurons precisely emulate the electrical activity of biological neurons, and consume less power, there are several obstacles faced when implementing them. First, the fabrication process is more complicated than the digital design based on normal cells and commercial tools. Second, the operation of analog neurons tends to vary with different process technologies, temperature, and voltage variations which introduces randomness in the neuron's function. Although it might be close to biological neuron, it is not easy to control. Furthermore, there is no simulation support for this design. These obstacles make prototyping and debugging large-scale analog neuromorphic systems difficult.

Table 9.1 Neuromorphic systems survey

Platform	Neurogrid	BrainScaleS	SpiNNaker	TrueNorth	Loihi
Technology	Analog, sub-threshold	Analog, over-threshold	Digital, programmable	Digital, fixed	Digital, programmable
Feature size	180 nm	180 nm	130 nm	28 nm	14 nm
Die size	1.7 cm^2	0.5 cm^2	1 cm^2	4.3 cm^2	60 mm^2
# transistor	23M	15M	100M	5.4B	2.07B
# neurons	65k	512	16k	1M	131k
# synapses	100M	100k	16M	256M	16-128M (8-1 bit/synapse)
Power	150 mW	1.3 W	1 W	72 mW	
Interconnect	Tree-multicast	Hierarchical	2D mesh-multicast	2D mesh unicast	2D mesh unicast
Neuron model	Adaptive quadratic IF	Adaptive exponential IF	Programmable	LIF	Programmable
Synapse model	Shared dendrite	4-bit digital	Programmable	Binary, 4 modes	1-8 bit
Run-time plasticity	No	STDP	Programmable	No	Programmable

9.2 Software Emulation Approach

The initial idea of operating neuromorphic systems comes from research in neuroscience which aim to mimic the operation of biological brains. For example, the Human Brain project [15] uses exascale supercomputers to investigate the brain on different spatial and temporal scales. SpiNNaker [11] is another project based on massively parallel processors to emulate the operation of biological brains.

There are also several software emulations of neuromorphic systems. Brian 2 [20] is a Python-based emulation tool for neuromorphic systems. Brain 2 relies on CPU performance, and its GPU version is still under development as of 2021 [21]. NEST [12] is another spiking neural network emulation platform that supports more than 50 neuron models. SpiNNaker team also released the software package for running PyNN simulation on SpiNNaker in [18]. Reusing the existing neural network platforms, PyCARL [4] and BindsNet [13] can use GPU to accelerate the emulation performance.

9.2.1 SpiNNaker

The SpiNNaker [11] project aims to develop a massively parallel supercomputer to emulate the operation of the biological brain in real-time. SpiNNaker consists of more than a million ARM9 processors and 7 Tbytes of distributed RAM throughout a 57K nodes system. Each node consists of 18 cores and 128 Mbytes SDRAM, and each core emulates a thousand neurons. Since SpiNNaker is based on general-purpose processors, its neuron and synapse models are not limited. However, unlike the supercomputer-based model, SpiNNaker relies on different computing and communication model that allows more efficient brain emulations.

Communication of SpiNNaker is based on a folded 2D triangular mesh. A router accepts packets from 18 resident cores and six incoming inter-chip links. The transmission uses the AER format to communicate between cores.

9.2.1.1 Model

SpiNNaker follows the "point neural model" (see Fig. 9.1) where the structure of dendrite is ignored. The inputs of the neuron are fed through to the *soma*. The learning mechanism only updates the *synapse*, which represents the strength of the connection. By going through the synapse, the input spikes will have different attitudes based on the *weight*. If the weight is positive, it is an excitatory synapse. Otherwise, it is an inhibitory synapse.

9.2.1.2 SpiNNaker Architecture

The architecture of the SpiNNaker node is shown in Fig. 9.2. It consists of 18 ARM processors connected via a system Network-on-Chip. The packet router receives six

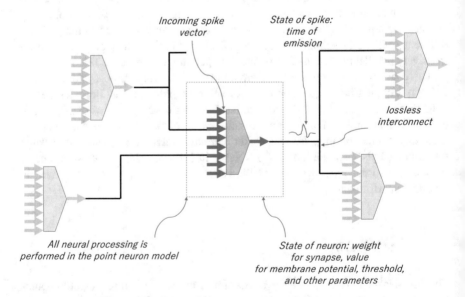

Fig. 9.1 The neuron point model

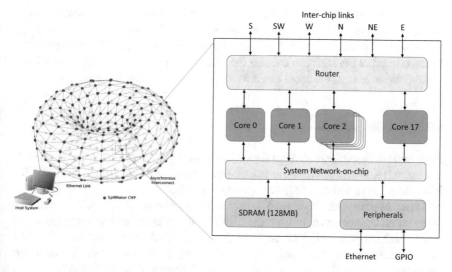

Fig. 9.2 The architecture of a SpiNNaker node

input links from neighboring nodes and one input link from one of 18 processors. In the packet router, packets are decoded, routed, and fed to one of the output six external links or one of the local processors. The node also supports internal RAM, ROM, Ethernet connection, controller, and external SRAM interface. By connecting the nodes, SpiNNaker forms a triangular lattice topology which is folded onto a toroid surface as shown in Fig. 9.2.

The human brain comprises of 10^{11} neurons. SpiNNaker aims to model 1% of this scale or ten whole mouse brains. Each neuron of the brain is supposed to connect with thousands of other neurons with the mean firing rate of 10 Hz and the maximal firing rate of hundreds of Hertz. In SpiNNaker, the target firing rate is at 10 Hz, and each neuron has 1000 inputs. With this model, the ARM process is designed to run at 200 MHz, and the router is to run at 100 MHz. These are reasonable frequencies for being balanced between biological performance and power consumption. With the interrupt-based wake-up manner, the process goes to sleep if there is no activity (i.e., no input spikes).

SpiNNaker uses ARM968E-S as its processor [2], which is part of the ARM 9family with 32 KB instruction and 64 KB data cache memory connected directly to the core. All parts of the core are mapped and accessible via the AHB (Advanced High-Speed Bus—one of the ARM AMBA—Advanced Microcontroller Bus Architecture—interface standards) interface. The process has direct access to the DMA (with system NoC to intra-node), communication (with communication NoC to inter-nodes), and other parts (interrupt, timer/counter). The data cache of the core can be accessed directly from the DMA controller which allows system NoC to deliver packets into the data cache memory.

In SpiNNaker, there is no conventional operating system running on the cores. Instead, it operates in an interrupt service manner. An interrupt arrives in a message form and wakes up the core to handle it. The packet first arrives in a packet buffer. Then, it triggers the DMA to transfer the result and update the synaptic inputs (the spikes). Besides that, the cores can also be woken up by external events.

9.2.1.3 Other Modules

Each node in SpiNNaker also has timers/counters. They are managed and accessible via the AHB interface. These counters/timers provide the real-time dynamics for biological spikes and other periodic timing controls.

Direct Memory Access is used to access the data cache of a core. This transfers inter-neural connection data from the SDRAM to the corresponding processor and writes back to update information (states, outputs). DMA also allows the host ARM core to access system NoC (including the SDRAM) in "bridge" mode. There is also an error management module with interruptions for recovering from crashes.

To connect with the host machine, SpiNNaker uses an Ethernet connection. Each node has an Ethernet interface, but few chips are connected to the host machine. The host machine runs the software used to program and monitor SpiNNaker. Instead, in the host machine, the software is run to program and monitor the SpiNNaker. As mentioned above, there is a cached memory within each processor. In particular, there are ITCM and DTCM for caching instruction and data. Both are run at 800 MBps speed and can be accessed at each core. The capacitance of ITCM and DTCM is 32 kB and 64 kB, respectively. There is a shared SRAM within the chip with a capacity of 32 kB and a speed of 25 MBps. SpiNNaker also has off-chip memory with 128MB SDRAM running at 64 MBps. Both SRAM and SDRAM are accessible at the node level. Since SpiNNaker is a 32-bit system, all memories are mapped into its 32-bit memory space. Consequently, the processors within a node can access this information. However, the nodes cannot access the information in the memory space from other nodes. Instead, packets delivered by 6-links interconnects are used to communicate. There are no such things as cache coherence between nodes.

9.2.1.4 Communication in SpiNNaker

To communicate between processors of a node, a system NoC is used. Since the memory space is node-level visible, processes can access the information it needs. For inter-node communication, packets are used. Packets are either in 40-bits (5 bytes) or 72-bit (9 bytes) format. In particular, a packet consists of 8 control bits and either one or two words. The second word is optional and indicates one bit in one of 8 control bits.

Packets are transmitted to one of the neighboring nodes. They will be routed to the local cores or propagated to one neighbor. Packets are either in uni-cast or multicast format.

The SpiNNaker system performs its inter-neuron communication based on *Address Event Representation* (AER) [19], where the address of neurons are sent instead of spikes. Compared to sending the vectors of the output of the neuron array, *AER* has lower complexity thanks to lesser bit and the sparsity of the firing events.

In SpiNNaker, the system consists of one million ARM cores. Each core emulates a thousand neurons. Consequently, there are one billion neurons in SpiNNaker. As each neuron has a different address in AER format, 32-bit is used to represent it. To send the spike, the system transmits the 32-bit ID of the neuron instead. This 32-bit ID is attached as the optional payload in SpiNNaker packets in Fig. 9.3. There are four basic types of packets:

- *Nearest Neighbor*: this type of packet is used to allow nodes to communicate with their neighbors. The main functions of this packet are: (1) initialize the system, (2) run-time flood-fill, and (3) debugging.
- *Point-to-Point*: this type of packet consists of the source node address and the destination node address (each one is 16-bit) in the first word. The router will use a look-up table to find the correct output link for routing the packet.

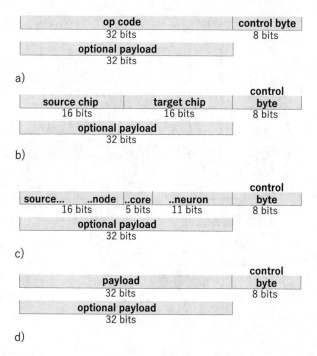

Fig. 9.3 The SpiNNaker packets. (**a**) Nearest neighbor packet layout. (**b**) Point-to-point packet layout. (**c**) The neural event multicast packet layout. (**d**) Fixed route packet layout

- *Multi-cast*: this type of packet is mainly used for inter-neural communication (i.e., spikes). The first word consists of the AER of the firing neuron in 32-bit. The router must use content-access-memory (CAM) to extract the proper direction for this type of packet.
- *Fixed-routing*: this type of packet already predefines the routing path in the content. This packet provides a fast track from the launching node to the Ethernet-enabled node.

There are two types of interconnect in SpiNNaker: system NoC and communication NoC. While the system NoC maintains the connectivity of 18 processors and the SDRAM, the communication NoC maintains the input and output packets. The communication NoC follows a tree-based structure where the input streams are merged into a single stream to be routed. Later, the output stream is demultiplexed into six streams to the neighbor nodes or to local. Figure 9.4 describes the flow of input stream. The inter-chip links are 4-bit each and running at 60 MHz. Meanwhile, the on-chip processor communication runs at 8-bit and 200 MHz. Note that there are 18 input streams from 18 processors. By merging all of them, the final stream is 72-bit at 100 MHz. This is one packet per cycle to be routed. SpiNNaKer also adopts the AER protocol as the central idea with its modifications to form the communications between neurons. As the AER format represents the spike in two information: (1)

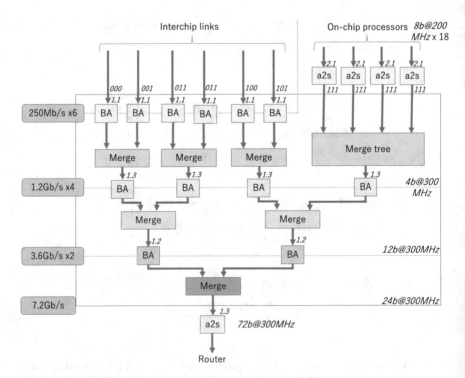

Fig. 9.4 The SpiNNaker input streams of the communication NoC

the time of spike and (2) the neuron's identity, SpiNNaker conveys the idea using packet-switch communication and broadcast/multicast routing. Once a spike occurs, the AER signal is sent to the packet-switch communication fabric, and delivered to the connected neurons.

By delivering by electronics communication fabrics, spikes can instantaneously arrive at their destinations compared to biological signals in the brain. This allows SpiNNaker to map any neuron to any node (cluster of neurons) and virtually form the biological topology regardless of the pack-switch communication topology. However, the problem of efficient mapping is also considerable and has been extensively investigated to reduce the transmission time between neurons [3].

Since the electronics signals can be delivered instantaneously, the "time" information of the spike cannot be preserved, and must be presented in another way. In the biological system, spikes are expected to be at the right place at the right time; however, the electronics system works the other way around. If there is no congestion in the communication infrastructure, the spikes are delivered effectively. However, congestion or failures in the communication system can delay, drop, or misroute spikes. Therefore, there is a need for a scheduling method and fault-tolerance features in the spiking system.

In SpiNNaker, there are fixed time-multiplexed 1000 neurons per ARM processor to be executed. Therefore, each node of SpiNNaker with 18 ARM processors, makes up 18,000 neurons per node. Therefore, SpiNNaker expects the delivery time (spike window) to be likely at $0.2\,\mu s$/hop to ensure the neurons react to the stimuli in order of ms like biological systems.

At first, the communication between ARM cores of a SpiNNaker chip is handled by a Network-on-Chip. Then, it is converted to off-chip communication using packet-router modules. Next, six links are merged using a time division multiplexer to stream together with the stream from the local NoC. The output stream is later split into six output links (Fig. 9.5).

The inter-node communication in SpiNNaker is via packets. The packets are generated by cores and transmitted to the local router. The packets are then redirected to the target cores. If the destination neurons are in the same node as the source, the local router sends the packets back to the local cores. If the destination is in other nodes, the local router sends the packets to a neighboring node. As each node can only connect to six nodes in general, routing techniques are needed to deliver the packets efficiently.

SpiNNaker packets are either in 40-bit or 72-bit format (8 control bit and either 4 bytes or 8 bytes of data). While the nearest neighbor packets are used to initialize the system, flood-fill communication (broadcast) and debugging, point-to-point packets allow more detailed communications. Among those packets, we focus on the *neural event multicast* as it represents the interneural communication. Here, the AER of SpiNNaker can be summarized in Fig. 9.3 where it uses 16 bits for node ID, 5 bit for the core ID (ARM core), and 11 bits for the neuron ID (neuron within ARM core). The routing method for AER follows the Content-Access-Memory (CAM) method, where it first looks up the CAM to find the address on the output RAM.

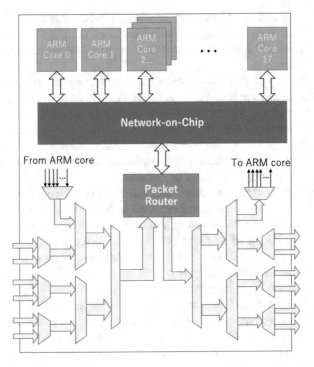

Fig. 9.5 The SpiNNaker node

Then, the output RAM of 6+18 encodes the output direct in one-hot format (the first 6 bits are for the inter-chip link and the last 18 bits for the internal core).

9.2.1.5 SpiNNaker Software Platform

The software platform of SpiNNaker is classified either as running in the system, or running in the host machine. For example, Fig. 9.6 depicts the software platform for SpiNNaker. The software running within SpiNNaker is mostly written in C. This software perform at a primitive level for both controlling and emulating the operation of the neuromorphic system.

Most SpiNNaker applications run based on an event management library named Spin1 API. This library provides a platform for associating the interrupts with event handling code. Without running any code, the processors will enter a low-power sleep mode. Once an event triggers an interrupt, the processor starts to operate.

The software running at the host machine is used to download, monitor, and visualize the data. This software communicates with SpiNNaker via the Ethernet connection as each node has one.

SpiNNaker provides APIs to allow designers to develop the SpiNNaer in either C, Perl, or Python at the host machine. In addition, some visualizers allow users to

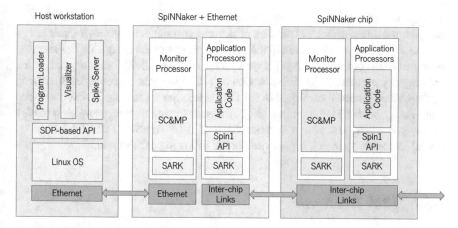

Fig. 9.6 The SpiNNaker software platform

obverse the input and output of the system (and neurons). The host machine can also supply input spike via a Spike Server application. This can be used to provide a data set or connect to input sensors that provide spiking inputs. The SpiNNaker has been demonstrated with Negro—a neuromorphic software platform, AER sensors, and robotics.

9.3   Digital Hardware Design Approach

While using general-purpose computing resources to emulate the operation of the neuromorphic system can be a feasible solution for neuroscience research, the energy efficiency of this method is not feasible for edge computing. For instance, the reference SpiNNaker system of 600 PCBs emulating 460 million neurons consumes 50 kW. To solve this problem, using hardware accelerators is considered a suitable approach.

The analog and mixed-signal approach is covered in the next section. Instead of using CPU or GPU to compute the neuron operation and synaptic plasticity, the digital hardware approach relies on the hardware design of neurons and synapses. Generally, spikes are sent in Address-Event-Representation format and decoded at the receiver to obtain the weighted spike. This synapse operation can be done using a resistive crossbar or a memory.

Due to the complexity of other neuron models, most digital hardware works use a variation of Integrated-and-Fire in their model (i.e., adding leaky or stochastic inputs). This allows implementation with low cost and high energy efficiency. However, the flexibility in neural computation is limited.

There are several works on digital hardware that have been done. Notably, TrueNorth by IBM [17] in 2014 and Loihi by Intel [7] in 2018. Both chips share the same principles: (1) neuron circuits serially emulate group of neurons; (2) neurons are grouped into clusters; (3) clusters of neurons communicate via an asynchronous 2D mesh network-on-chip. Due to the low complexity of the design, online learning becomes a new challenge. In [9], a neuromorphic chip with 256 neurons and online learning is presented. An updated version in [10] illustrates four cores of Stochastic Spike-Driven Online Learning. NASH [14] is another multi-core 3D-Network-on-Chip design with STDP learning. In [22], a neuro-inspired computing chip survey is also presented. The design and operation of IBM TrueNorth and Intel Loihi as digital hardware approach is presented in the following subsections.

9.3.1   IBM TrueNorth

IBM TrueNorth [17] targets to deliver a dense and energy-efficient platform for cognitive applications. The TrueNorth chip consists of 5.4 million transistors fabricated in 28 nm CMOS technology. Each chip houses 4096 neurosynaptic cores that communicate in 2D Mesh 64×64 asynchronous Network-on-Chip. Each core consists of 256 neurons and 256 inputs which make 65,536 synapses. The chip is asynchronous with a 1 kHz time-step clock [1]. Integrate and Fire (IF) neuron model is adopted together with 1-bit synapse and neuron-based fixed weight and 23 configurable parameters. In each core, only one physical IF neuron is used to perform 256 neuron operations in a serial manner. TrueNorth is trained off-line and the chip only performs inference. Software for modeling, training, and mapping the desired system into the TrueNorth system is available.

9.3.1.1   TrueNorth Neurosynaptic Architecture

TrueNorth chips can be connected to form larger systems. For example, the NS16e circuit board hosts 16 IBM TrueNorth chips in a layout of 4×4. The 2D Mesh Network-on-Chip is naturally extended by connecting TrueNorth chip is 2D Mesh layouts. In TrueNorth, packets are routed by the offset between two nodes (difference in X and Y coordinates). Therefore, theoretically, the TrueNorth can efficiently expand the size of NoC. TrueNorth architecture is based on asynchronous 2D Network-on-Chip. The chip layout and inter-chip communication is shown in Fig. 9.7 [1]. The neuron synaptic core (see Fig. 9.8) which consists of 256 neurons is attached to a 2D-NoC router. The spikes are sent from the pre-synaptic neurons to the network, where the routers help deliver them to the destination neuro synaptic core where the post-synaptic core resides. The spikes in AER format can travel through chip periphery to off-chip wires and delivered to other chips.

Figure 9.8 depicts the model for a single neuro synaptic core of TrueNorth. The input spikes arrive at axons. The spikes are sent to the neuron via a dendrite. The connection between axons and dendrites is the synapses. In TrueNorth core, these connections are represented in a synaptic crossbar where black dots connect

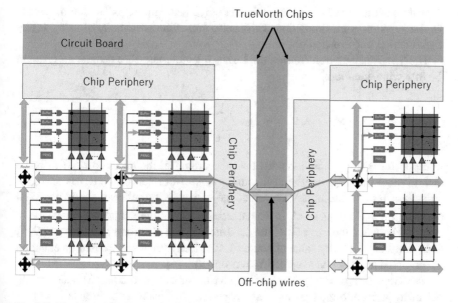

Fig. 9.7   The inter-chip communication

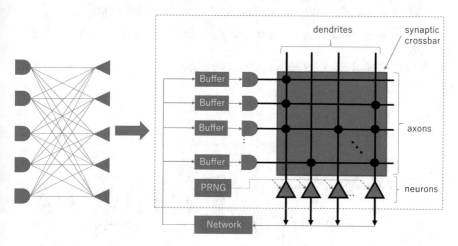

Fig. 9.8   The neural network model in the neurosynaptic core [1]

synapses. There are only two modes in the synapses: connected and unconnected, allowing the weight to be represented in binary. The output of the synaptic crossbar is later multiplied with a fixed value; therefore, the actual weights are not binary but fixed for each neuron (0 or a fixed value).

The neuron in TrueNorth operates in a synchronized 1 kHz global clocks. Meanwhile, the NoC is asynchronous, and the cores operate independently. This means all computation of neuron must be done within a time frame of 1 ms regardless of the deviation in arrival time of the spikes. This helps TrueNorth operate similar to its software simulation. This is also one of the advantages of using a fully digital neural implementation.

The TrueNorth neuron follows the Leaky Integrate-and-Fire fashion an given in the following equation:

$$V_j(t) = V_j(t-1) + \sum_{i=0}^{N-1} A_i(t)w_{i,j}s_j^{G_i} - \lambda_j \tag{9.1}$$

where $V_j(t)$ is the membrane potential of neuron j at the timestep t, $A_i(t)$ is the binary value that represents the incoming spike from neuron i (1: has spike, 0: no spike), $w_{i,j}$ is the binary represent the connectivity between the axon i and the dendrite j (1: connected, 0: unconnected), $s_j^{G_i}$ is the actual weight (synapse strength) between the axon i and the dendrite j and is classified in four type (G_i: 1 to 4), and λ_j is the leaky value of neuron j. The leaky value and the neuron equation are not fixed and may be adapted in several modes [6].

The TrueNorth processor supports stochastic spike integration, leak, and threshold using a pseudo-random number generator. The leaky and reset is also programmable. The architecture of the neuron is depicted in Fig. 9.9.

The operation of a neuron are summarized in the following steps:

- The neuro synaptic core receives spikes from the NoC and stores the spikes in buffers.
- A 1 kHz global clock comes and triggers the operation of all neuro synaptic core.

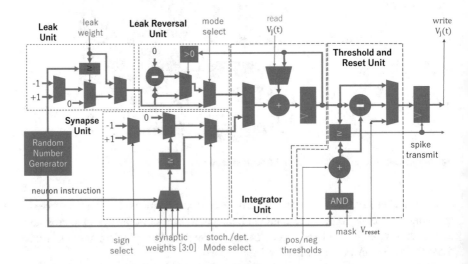

Fig. 9.9 The architecture of TrueNorth neuron [1]

- The spikes ($A_i(t)$) are sent from the buffers through the dendrites as depicted in Fig. 9.8.
- The neuron receives the spikes ($A_i(t)w_{i,j}$) and follows the Eq. (9.1).
- At the end of the time step, if the membrane potential exceeds the threshold, a spike is generated and sent into the network.
- After firing, the membrane potential of the neuron is reset to a resting voltage.

As shown in Fig. 9.9, the synaptic weight is compared with a random number in stochastic mode. This inserts randomness into the neuron model. We also can observe the random leak mode and random in threshold comparison. In the deterministic model, the operation of the neuron follows the simplest LIF.

In each neuro synaptic core of TrueNorth, there is only one physical neuron used for simulating 256 neurons in serial fashion. Each neuron parameters and synaptic connection (410-bit format) is loaded for computation consequently. The main reason is the system target to run in biological real-time with a 1 kHz global synchronization clock which does not need to perform neural computation in a parallel manner. This has been the theme for digital neuron design as its weights and parameters can be stored and reloaded.

TrueNorth Synapses Design As mentioned above, the synaptic crossbar of the neuron stores connection in a binary format (1: connected, 0: unconnected). The synaptic connections are embedded into a single SRAM of 256 rows by 410 columns as the neuron operates serially. Each row consists of 256 bits of synaptic connection, 124 bits of neuron parameters, 26 bits of spike destination (18-bit for network routing and 8-bit for axon index), and 4 bits of spike delivery tick.

The input spikes in 8-bit AER format are used as the index for the synaptic connection to obtain the connection status (0/1: unconnected/connected). If the axon and the neurons are not connected, the return weight is 0, which means no incoming weighted spike.

9.3.1.2   Interconnect

Once a neuron fires, the spike is packaged and sent to the network to deliver to a destination core. Since neurons are clusterized into neuro synaptic core with 256 neurons/core, the inter-core connection is required. TrueNorth is based on a unicast-only 2D Mesh asynchronous network. Therefore, the architecture can be tiled for not only the core-level but also at the chip-level as depicted in Fig. 9.7. By exploiting the scalability of 2D Mesh topology, The TrueNorth chips are placed in a 2D mesh in the PCB. The off-chip wires help connect two chips and then extend the on-chip NoC to the new one. This continues the communication of the asynchronous 2D NoC to the new chip.

The routing process at each router consists of six phases: (1) from local, (2) forward east, (3) forward west, (4) forward north, (5) forward south, and (6) to local. To route the packet, 18-bit is used for dx and dy, which are the number of hops in the x- and y-direction. For instance, a packet from node (1,3) to node (4,5) (format:

(dy, dx)) has the initial value of dy and dx are -3 and -2, respectively. Since the priority of routing is east (increase dx), west (decrease dx), north (increase dy), and south (decrease dy), the packet will travel to the east first to the node (1,4). By routing to the east direction, the dx value is increased by one. At node (1,4), the value of dy and dx are -3 and -1, respectively. Once both dx and dy gets to zero, the packet arrives at the destination.

If more than one messages compete for the same router, a first-come-first-serve basis is adopted to service the routing processes.

Since it is a unicast design, multicast packets are handled by repetitively sending the packets to a list of destinations. The repetition is done by cloning the neuron within the same core (node).

9.3.1.3   TrueNorth Software

The hardware of TrueNorth is supported by a software emulator, which can provide a deterministic behavior as in hardware. Therefore, application development can be performed in software and later mapped into hardware. Likewise, the training is conducted in software and later mapped into TrueNorth chips.

The software model for TrueNorth is based on modeling a network consisting of neuro synaptic cores, which are non-divisible building block [8]. TrueNorth team develops a new programming paradigm called *Corelet* which has several parts: (1) a corelet that is an abstraction model encapsulating the detail of neurosynaptic cores; (2) Corelet language, which is an object orient programing language to create, compose and decompose Corelet; (3) a Corelet library for reusing the program and (4) a programing environment that consists of Compass—the TrueNorth architectural simulator. The simulator provides a one-to-one result to the TrueNorth system.

Corelet abstraction model consists of *corelet seed* which is a single neurosynaptic core exposing only input and output while encapsulating under-the-hood the parameters and operation. The programming only accesses the input and output of the *corelet seed*. By connecting two or more corelet seeds, we end up with a corelet. Several corelets can be connected to obtain a new corelet. From a programming perspective, the input and output of the Corelet will be connected for modelling [8]. For programming, *Corelet Language* provides corelet classes that allow designers to model, program, and simulate in the Compass simulator. Figure 9.10 depicts the corelet seed and the composition of corelet. Figure 9.10a illustrate the corelet which consists of neuron and input/output connectors. The corelet seed encapsulates the intra-core connection and only exposes the external connection to obtained a block in Fig. 9.10b. If there are two corelets connected, they can be wrapped into a single one. the connection between two corelet A and B is shown in Fig. 9.10c. Then, Corelet C wraps the two Corelet A and B. Corelet C in the programming model, which encapsulates the two internal corelets and connections and only exposes the inputs and outputs.

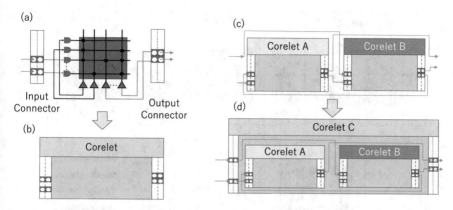

Fig. 9.10 The TrueNorth corelet framework. (**a**) Neurosynaptic core with input and output connector. (**b**) Corelet provides a wrapped model which only exposes the input-output. (**c**) Two connected corelet A and B. (**d**) Corelet C wraps two corelet A and B and now expose only the input and output connectors

The Compass simulator is written in C++ with MPI library call and OpenMP threading primitives. Compass has been scaled to run the network with 530 billion neurons and 137 trillion synapses.

The applications for TrueNorth have been demonstrated at remarkable low power level [17] such as speaker recognition, composer recognition, digit recognition, eye detection, and multi-object recognition.

9.3.2   Intel Loihi

Loihi is a 60 mm^2 neuromorphic chip fabricated in 14 nm process in 2018. The overall structure of Loihi is based on 2D-mesh NoC with 128 neuromorphic cores and three embedded x86 cores per chip. Loihi can be scaled up to 4096 cores and support up to 16,384 inter-chips communication. Each core implements 1024 neural units with variable synaptic resolution. The primitive spiking neural units are grouped into sets of trees. Each core also supports a programmable learning engine. Learning rules are microcode programmable and support features on input and output synaptics. Learning on Loihi supports simple pairwise STDP and more complicated ones such as triplet STDP or reinforcement learning.

Figure 9.11 depicts the top-level architecture of Loihi. The incoming spike has information on the axon index, and the synapse unit processes and maps the incoming spike to the synaptic weight from memory. In contrast, the dendrite unit updates the neuron's internal values. Axon unit generates the output spike based on the membrane potential. Finally, the learning unit updates the synaptic weight based on the dendrite output and the weight in the synapse. The weight and neuron in the

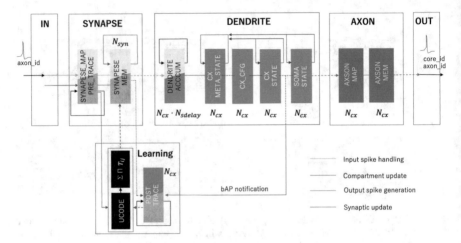

Fig. 9.11 Top-level architecture of Loihi

Loihi core are variable (1–8 bit). Therefore, the system can be programmed to have different densities and resolutions.

The interconnect for Loihi is a 2D asynchronous Network-on-Chip with the ability to expand multi-chip due to the scalability of 2D mesh. Once a neuron fires, a fan-out structure at the axon finds and sends the spikes to all downstream neurons. The downstream neuron might belong to the current core or others in the network. In Loihi, the maximal number of neurons per core is 1024, and the maximal core-to-core fan-out must not exceed 4096.

9.4   Analog and Mixed-Signal Hardware Approach

The original neuromorphic system designed by Carver Mead and his team has become the standard for designing and implementing neuromorphic systems in the analog and mixed signal approach. Instead of emulating the function of neuron digitally with a certain level of granularity, analog neurons allow the actual operation of biological neurons to be mimicked.

9.4.1   NeuroGrid

Neurogrid from Stanford University consists of two parts: software to visualize the neuromorphic system and hardware to perform real-time simulation. The software of Neurogrid composes of a user interface (UI), hardware abstraction layer (HAL), and driver for the hardware. Figure 9.12 depicts the overall system of Neurogrid.

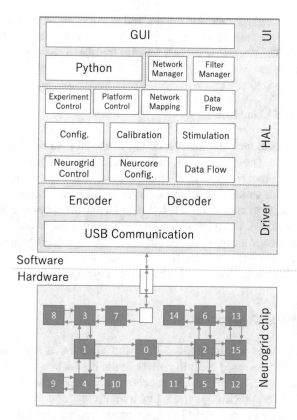

Fig. 9.12 Neurogrid system with both hardware and software [5]

While the user interface provides the programming model for the neural network, the hardware abstraction layer maps the model into the Neurogrid circuit. Finally, the driver helps download and deploy the model over USB into the Neurogrid system.

9.4.1.1 Neurogrid Software

As depicted in Fig. 9.12, the software of Neurogrid consists of three major parts: UI, HAL, and Driver. The UI is the graphic interface with NGPython allows users to specify the neuron models in the Python programming environment. It also provides the control interface and visualization of the results.

In *HAL*, the network and filter management connects the GUI with the simulation platform to provide the network's connectivity and activity. The network mapping converts the Python environment into router configuration. Data flow translate from the model to hardware.

In *Driver*, the configuration for routers and neurons are generated. Then, the Neurogrid packets are created by data flow and encoded to be sent to Neurogrid via USB connection. The Neurogrid packets are also translated back to normal data for visualization purposes.

9.4.1.2 Neurogrid Hardware

Neurogrid uses a sub-threshold analog circuit to model neurons and synapses in biological real-time. The spikes are digitally communicated using on-chip networks as analog spikes cannot travel long-distance without losses. A Neurogrid packet is a sequence of 12-bit words consisting of the route, address, payload, and tail. The route information instructs a core to forward to the next node or consume it (as it reaches the destination). The payload has several types of information: a spike index for the row in the neural array, data to be written to RAM, an analog signal converted by one of four ADCs, and a tail filed to end the packet.

The on-chip network of Neurogrid is tree-based topology as depicted in Fig. 9.12.

9.4.1.3 Neurogrid Communication

The Neurogrid spikes are dispatched from the array by a transmitter. Spikes travel the on-chip tree-based NoC to deliver at the destination by a receiver. The communication system is a digital circuit operating asynchronously.

Spikes from a neuron will be selected by a transmitter and sent to the network. The tree NoC router of Neurogrid has three ports: top left and right. If the post-synaptic neurons are on the right or left sides, it travels from the source to the router to the dedicated port (right or left). If the post-synaptic neurons are on a different branch, spikes are sent to the top and later distributed into either the left or right units.

The route field of the packet is stored at the most significant bits and later shifted. If the stop code (all zeros) has remained, the spikes are distributed in a multicast manner (in both left and right and local). For example, Fig. 9.13 depicts the case

Fig. 9.13 Neurogrid packet transmission

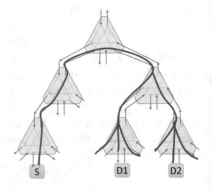

when the neurons in the source node (S) desires to send spikes to neurons in two destination nodes (D1 and D2). Since D1 and D2 belong to a different branch, the packet is routed to the top port twice first. Then, it takes a right turn. When it encounters the stop code, the packet is multicasted to all branches.

9.4.1.4 Transmitter and Receiver

Figure 9.14 shows the transmitter and receiver architecture. In the transmitter, interface **I** receives and relays the request from spiking neuron S to the hierarchical row arbiters (J). There is another interface and arbiter in the column to select the spiking neuron spikes. The spike packet consists of column address (X), row address (Y), and a tail word concluding the packet. If the packet is acknowledged, continue by forwarding the sequencer (SEQ).

As depicted in Fig. 9.14b, the receiver receives the packet consisting of the row and column information. The row and column information is checked to see whether it fits the row and column index. Once it delivers at the correct row and column, it forwards to the spiking neuron (S) and issues the acknowledgment (ACK).

The router in Neurogrid has two phases of operation. The first phase is *point-to-point* where the router route the packet up/down or left/right based on the first word's information of the packet. The second phase is *branching* where the router duplicates the packet to both the left and right ports for multicast routing. The duplicated packets can be received by the local neuron array or being filtered using information from a local SRAM of the core.

The NoC of Neurogrid is organized in tree-based topology, which is fit for multicasting. In the first phase, packets are delivered to the destination root router.

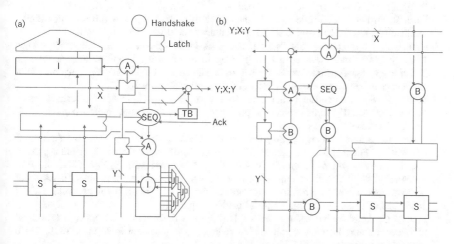

Fig. 9.14 Neurogrid transmitter and receiver architecture [5]. (**a**) Transmitter. (**b**) Receiver

Then, the packets are distributed into branches regardless of having receipting neuron. As the packets are later filtered, it can reduce the routing due to low look-up table complexity.

9.5   Chapter Summary

This chapter presented a comprehensive survey of research works in neuromorphic computing systems. These works are categorized as software emulation, digital hardware, and analog and mixed-signal hardware approaches. This chapter provided an exhaustive review of the research conducted in neuromorphic computing and illuminated the gaps in the field where new research is needed.

References

1. Akopyan F, Sawada J, Cassidy A, Alvarez-Icaza R, Arthur J, Merolla P, Imam N, Nakamura Y, Datta P, Nam G-J et al (2015) Truenorth: Design and tool flow of a 65 mw 1 million neuron programmable neurosynaptic chip. IEEE Trans Comput Aided Des Integr Circuits Syst 34(10):1537–1557
2. ARM (2004) ARM968E-S technical reference manual. ARM, arm ddi 0311c edition
3. Balaji A, Das A, Wu Y, Huynh K, Dell'Anna FG, Indiveri G, Krichmar JL, Dutt ND, Schaafsma S, Catthoor F (2019) Mapping spiking neural networks to neuromorphic hardware. IEEE Trans Very Large Scale Integr Syst 28(1):76–86
4. Balaji A, Adiraju P, Kashyap HJ, Das A, Krichmar JL, Dutt ND, Catthoor F (2020) PyCARL: a PyNN interface for hardware-software co-simulation of spiking neural network. Preprint, arXiv:2003.09696
5. Benjamin BV, Gao P, McQuinn E, Choudhary S, Chandrasekaran AR, Bussat J-M, Alvarez-Icaza R, Arthur JV, Merolla PA, Boahen K (2014) Neurogrid: A mixed-analog-digital multichip system for large-scale neural simulations. Proc IEEE 102(5):699–716
6. Cassidy AS, Merolla P, Arthur JV, Esser SK, Jackson B, Alvarez-Icaza R, Datta P, Sawada J, Wong TM, Feldman V et al (2013) Cognitive computing building block: a versatile and efficient digital neuron model for neurosynaptic cores. In: The 2013 international joint conference on neural networks (IJCNN). IEEE, Piscataway, pp 1–10
7. Davies M, Srinivasa N, Lin T-H, Chinya G, Cao Y, Choday SH, Dimou G, Joshi P, Imam N, Jain S et al (2018) Loihi: a neuromorphic manycore processor with on-chip learning. IEEE Micro 38(1):82–99
8. Esser SK, Andreopoulos A, Appuswamy R, Datta P, Barch D, Amir A, Arthur JV, Cassidy A, Flickner M, Merolla P, Chandra S, Basilico N, Carpin S, Zimmerman T, Zee F, Alvarez-Icaza R, Kusnitz JA, Wong TM, Risk WP, McQuinn E, Nayak TK, Singh R, Modha DS (2013) Cognitive computing systems: algorithms and applications for networks of neurosynaptic cores. In: The 2013 international joint conference on neural networks, IJCNN 2013, Dallas, TX, 4–9 Aug 2013. IEEE, Piscataway, pp 1–10
9. Frenkel C, Lefebvre M, Legat J-D, Bol D (2018) A 0.086-mm2 12.7-pj/sop 64k-synapse 256-neuron online-learning digital spiking neuromorphic processor in 28-nm CMOS. IEEE Trans Biomed Circuits Syst 13(1):145–158
10. Frenkel C, Legat J-D, Bol D (2019) Morphic: a 65-nm 738k-synapse/mm2 quad-core binary-weight digital neuromorphic processor with stochastic spike-driven online learning. IEEE Trans Biomed Circuits Syst 13(5):999–1010

11. Furber SB, Galluppi F, Temple S, Plana LA (2014) The spinnaker project. Proc IEEE 102(5):652–665
12. Gewaltig M-O, Diesmann M (2007) Nest (neural simulation tool). Scholarpedia 2(4):1430
13. Hazan H, Saunders DJ, Khan H, Patel D, Sanghavi DT, Siegelmann HT, Kozma R (2018) Bindsnet: a machine learning-oriented spiking neural networks library in python. Front Neuroinform 12:89
14. Ikechukwu OM, Dang KN, Abdallah AB (2021) On the design of a fault-tolerant scalable three dimensional NoC-based digital neuromorphic system with on-chip learning. IEEE Access 9:64331–64345
15. Markram H, Meier K, Lippert T, Grillner S, Frackowiak R, Dehaene S, Knoll A, Sompolinsky H, Verstreken K, DeFelipe J et al (2011) Introducing the human brain project. Procedia Comput Sci 7:39–42
16. Mead C (1989) Analog VLSI and neural systems. Addison-Wesley Longman, Reading
17. Merolla PA, Arthur JV, Alvarez-Icaza R, Cassidy AS, Sawada J, Akopyan F, Jackson BL, Imam N, Guo C, Nakamura Y et al (2014) A million spiking-neuron integrated circuit with a scalable communication network and interface. Science 345(6197):668–673
18. Rhodes O, Bogdan PA, Brenninkmeijer C, Davidson S, Fellows D, Gait A, Lester DR, Mikaitis M, Plana LA, Rowley AGD, Stokes AB, Furber SB (2018) sPyNNaker: a software package for running PyNN simulations on spinnaker. Front Neurosci 12:816
19. Sivilotti MA (1991) Wiring considerations in analog VLSI systems, with application to field-programmable networks. PhD thesis, California Institute of Technology
20. Stimberg M, Brette R, Goodman DF (2019) Brian 2, an intuitive and efficient neural simulator. eLife 8:e47314
21. Stimberg M, Goodman DF, Nowotny T (2020) Brian2genn: accelerating spiking neural network simulations with graphics hardware. Sci Rep 10(1):1–12
22. Zhang W, Gao B, Tang J, Yao P, Yu S, Chang M-F, Yoo H-J, Qian H, Wu H (2020) Neuro-inspired computing chips. Nat Electronics 3(7):371–382

Index